ENVIRONMENTAL SCIENCE, ENGINEERING AND TECHNOLOGY

STRATEGIC ADVANCES IN ENVIRONMENTAL IMPACT ASSESSMENT

CHALLENGES OF UNCONVENTIONAL SHALE GAS EXTRACTION

ENVIRONMENTAL SCIENCE, ENGINEERING AND TECHNOLOGY

Additional books and e-books in this series can be found
on Nova's website under the Series tab.

Environmental Science, Engineering and Technology

Strategic Advances in Environmental Impact Assessment

Challenges of Unconventional Shale Gas Extraction

Afsoon Moatari-Kazerouni

Copyright © 2019 by Nova Science Publishers, Inc.

All rights reserved. No part of this book may be reproduced, stored in a retrieval system or transmitted in any form or by any means: electronic, electrostatic, magnetic, tape, mechanical photocopying, recording or otherwise without the written permission of the Publisher.

We have partnered with Copyright Clearance Center to make it easy for you to obtain permissions to reuse content from this publication. Simply navigate to this publication's page on Nova's website and locate the "Get Permission" button below the title description. This button is linked directly to the title's permission page on copyright.com. Alternatively, you can visit copyright.com and search by title, ISBN, or ISSN.

For further questions about using the service on copyright.com, please contact:
Copyright Clearance Center
Phone: +1-(978) 750-8400   Fax: +1-(978) 750-4470   E-mail: info@copyright.com.

NOTICE TO THE READER

The Publisher has taken reasonable care in the preparation of this book, but makes no expressed or implied warranty of any kind and assumes no responsibility for any errors or omissions. No liability is assumed for incidental or consequential damages in connection with or arising out of information contained in this book. The Publisher shall not be liable for any special, consequential, or exemplary damages resulting, in whole or in part, from the readers' use of, or reliance upon, this material. Any parts of this book based on government reports are so indicated and copyright is claimed for those parts to the extent applicable to compilations of such works.

Independent verification should be sought for any data, advice or recommendations contained in this book. In addition, no responsibility is assumed by the publisher for any injury and/or damage to persons or property arising from any methods, products, instructions, ideas or otherwise contained in this publication.

This publication is designed to provide accurate and authoritative information with regard to the subject matter covered herein. It is sold with the clear understanding that the Publisher is not engaged in rendering legal or any other professional services. If legal or any other expert assistance is required, the services of a competent person should be sought. FROM A DECLARATION OF PARTICIPANTS JOINTLY ADOPTED BY A COMMITTEE OF THE AMERICAN BAR ASSOCIATION AND A COMMITTEE OF PUBLISHERS.

Additional color graphics may be available in the e-book version of this book.

Library of Congress Cataloging-in-Publication Data

ISBN: 978-1-53614-433-8

Published by Nova Science Publishers, Inc. † New York

CONTENTS

Executive Summary		**vii**
Chapter 1	Introduction	**1**
Chapter 2	Shale Gas Development in the Context: What Is Shale Gas?	**17**
Chapter 3	Shale Gas Technology and Well Integrity	**23**
Chapter 4	Water	**49**
Chapter 5	Greenhouse Gases and Other Air Emissions	**85**
Chapter 6	Land and Seismic Impacts	**101**
Chapter 7	Human Health	**115**
Chapter 8	Monitoring and Research	**129**
Chapter 9	Management and Mitigation	**169**
Chapter 10	Conclusion	**197**
Glossary		**205**
References		**211**

About the Author	**241**
Index	**243**

EXECUTIVE SUMMARY

Shale gas is natural gas that is tightly locked within low permeability sedimentary rock. Recent technological advances are making shale gas reserves increasingly accessible and their recovery more economically feasible. This resource is already being exploited in South Africa, China, the United States and Canada. Shale gas is being produced in large volumes, and will likely be developed in coming years on every continent except Antarctica. Depending on factors such as future natural gas prices and government regulations, further development of shale gas resources could potentially span many decades and involve the drilling of tens of thousands of hydraulically fractured horizontal wells.

This development is changing long-held expectations about oil and gas resource availability; several observers have characterized it as a game changer. Abundant, close to major markets, and relatively inexpensive to produce, shale gas represents a major new source of fossil energy. However, the rapid expansion of shale gas development over the past decade has occurred without a corresponding investment in monitoring and research addressing the impacts on the environment, public health, and communities. The primary concerns are the degradation of the quality of groundwater and surface water (including the safe disposal of large volumes of wastewater); the risk of increased greenhouse gas (GHG) emissions (including fugitive methane emissions during and after production), thus exacerbating anthropogenic climate change; disruptive effects on communities and land; and adverse effects on to human health. Other concerns include the local release of air contaminants and the potential for triggering small- to moderate-sized earthquakes in seismically active areas. These concerns will vary by region. The shale gas regions can be found near urban areas, presenting a large diversity in their geology, hydrology, land uses, and population density. The phrase environmental impacts from shale gas development masks many regional differences that are essential to understanding these impacts.

What is the state of knowledge of potential environmental impacts from the exploration, extraction, and development of shale gas resources, and what is the state of knowledge of associated mitigation options?

The assessment of environmental impacts is hampered by a lack of information about many key issues, particularly the problem of fluids escaping from incompletely sealed wells. If wells can be sealed, the risk to groundwater is expected to be minimal, although little is known about the mobility and fate of hydraulic fracturing chemicals and wastewater in the subsurface. The pertinent questions are difficult to answer objectively and scientifically, either because the relevant data have not been obtained or because some relevant data are not publicly available; or because existing data are of variable quality, allow for divergent interpretations, or span a wide range of values with different implications.

Two issues of particular concern are water resources, especially groundwater, and GHG emissions, both relate to well integrity. Many of the operational procedures used in shale gas extraction are similar to those used in conventional oil and gas extraction. Thus, industry experience is relevant to understanding these issues.

Natural gas leakage from improperly formed, damaged, or deteriorated cement seals is a long-recognized yet unresolved problem that continues to challenge engineers. Leaky wells due to improperly placed cement seals, damage from repeated fracturing treatments, or cement deterioration over time, have the potential to create pathways for contamination of groundwater resources and to increase GHG emissions. The issue of well integrity applies to all well types, including water and conventional gas or oil wells. Several factors make the long-term impact related to leakage greater for shale gas development than for conventional oil and gas development. These are the larger number of wells needed for shale gas extraction; the diverse chemicals used in hydraulic fracturing operations; the potential development of shale gas resources in rural and suburban areas that rely on groundwater resources; and possibly the repetitive fracturing process itself.

ENVIRONMENTAL IMPACTS

Water

Accidental surface releases of fracturing chemicals and wastewater, and changes in hydrology and water infiltration caused by new infrastructure, may affect shallow groundwater and surface water resources. A risk to potable groundwater exists from the upward migration of natural gas and saline waters from leaky well casings, and possibly natural fractures in the rock, old abandoned wells, and permeable faults. These pathways may allow for migration of gases and possibly saline fluids over long time scales, with potentially substantial cumulative impact on aquifer water quality. The risks due to surface activities will likely be minimal if proper precautionary management practices are followed. However, not enough is known about the fate of the chemicals in the flowback water to understand potential impacts to human health, the environment, or to develop appropriate remediation. Monitoring, assessment, and mitigation of impacts from upward migration are more difficult than for surface activities.

The greatest threat to groundwater is gas leakage from wells for which even existing best practices cannot assure long-term prevention. The degree to which natural assimilation capacity can limit the impacts of well leakage is site-specific due to variability in the magnitude of natural gas fluxes (or loadings) and aquifer hydro-geochemical compositions. These potential impacts are not being systematically monitored, predications remain unreliable, and approaches for effective and consistent monitoring need to be developed.

On average, about one-quarter to half of the water used in a single hydraulic fracturing treatment returns up the well to the surface after stimulation. This return flow, or flowback, is a potentially hazardous waste because it typically contains hydrocarbons including variable amounts of benzene and other aromatics, fracturing chemicals, and potentially hazardous constituents leached from the shale (e.g., salts, metals, metalloids, and natural radioactive constituents). Although flowback water is now commonly re-used in later fracturing treatments, a fraction eventually remains that poses technical challenges for treatment where deep wastewater injection for disposal may not be feasible.

Greenhouse Gas Emissions

To the extent that natural gas extracted from shale replaces oil and coal in energy use, particularly in electricity generation, it may reduce the environmental impact of fossil fuels and help to slow anthropogenic climate change. Whether shale gas development will actually reduce GHG emissions and slow climate change will depend on several variables, including which energy sources it displaces (*viz.,* coal and oil vs. nuclear and renewables), and the volume of methane emissions from gas leakage at the wellhead and in the distribution system. Experts disagree about these matters. Some conclude that downstream GHG benefits may be offset by upstream leakage, as well as the risk that gas undercuts the markets for lower carbon alternatives and fosters lock-in to high carbon infrastructure. Others argue that shale gas could provide a bridge to a low-carbon future. Furthermore, fields that produce gas with high carbon dioxide content could become an important additional source of carbon dioxide emissions unless the carbon dioxide is captured and used for enhanced oil recovery or is sequestered in saline aquifers.

Other Impacts

Land

Large-scale shale gas development may represent the start of several decades of production and the drilling of tens of thousands of wells. This development will have both local and dispersed land effects. The assessment of the environmental effects of shale gas development cannot, therefore, focus on a single well or well pad, but must also consider regional and cumulative effects.

Shale gas development requires extensive infrastructure that includes roads, well pads, compressor stations, pipeline rights-of-way, and staging areas. While the use of multi-well pads and longer horizontal laterals reduces the environmental impact, compared to individual well sites, the cumulative effects of the large number of wells and related infrastructure required to develop the resource still impose substantial impacts on communities and ecosystems. Furthermore, the performance of the infrastructure, operations, and closure procedures will likely be geology- and operator-specific and require monitoring for potential fluid migration over long time scales to assess impacts. Since the degree of future land reclamation from shale gas development is uncertain, consideration should be given to the risks and financial liability that arise. Land impacts may include deforestation, the destruction and fragmentation of wildlife habitat, and adverse effects on existing land uses such as agriculture and tourism. It is difficult to estimate these impacts without information on the location, pace, and scale of future shale gas development.

Human Health and Social Impacts

The health and social impacts of shale gas development have not been well studied. While shale gas development will provide varied economic benefits, it may also adversely affect water and air quality and community well-being as a result of the rapid growth of an extraction industry in rural and semi-rural areas. Potential community impacts include health and safety issues related to truck traffic and the sudden influx of a large transient workforce. Psychosocial impacts on individuals and on the communities have been reported related to physical stressors, such as noise, and perceived lack of trustworthiness of the industry and government. If shale gas development expands, risks to quality of life and well-being in some communities may become significant due to the combination of diverse factors related to land use, water quality, air quality, and loss of rural serenity, among others. These factors are particularly relevant to the ability to maintain their traditional way of life; concerns about the possible impacts of shale gas development on their quality of life and their rights

Air Contaminants

The emission of air pollutants from shale gas development is similar to conventional gas, but higher per unit of gas produced because of the greater effort required. These pollutants include diesel-use emissions, hydrocarbons, volatile organic compounds (e.g., benzene), and particulate matter. The main regional air emission issue is the generation of ozone, which in some circumstances could adversely affect air quality.

Seismic Events

Although hydraulic fracturing operations can cause minor earthquakes, most of the earthquakes that have been felt by the public have been caused not by the hydraulic fracturing itself, but by wastewater re-injection. Most experts judge the risk of hydraulic fracturing causing earthquakes to be low. Micro-seismic monitoring during operations can diminish this risk further. The risk by injection of waste fluids is greater but still low, and can be minimized through careful site selection, monitoring and management.

PUBLIC ACCEPTABILITY

The potential impacts of shale gas development, as well as strategies to manage these impacts, need to be considered in the context of local concerns and values. More specifically, the manner in which residents are engaged in decisions concerning shale gas development will be an important determinant of their acceptance or rejection of this development. To earn public trust, credible multidisciplinary research will need to be conducted to understand existing impacts and predict future impacts. Public acceptance of large-scale shale gas development will not be gained through industry claims of technological prowess or through government assurances that environmental effects are acceptable. It will be gained by transparent and credible monitoring of the environmental impacts.

LIMITS TO OUR KNOWLEDGE AND UNDERSTANDING

The technologies used by the shale gas industry have developed incrementally over several decades. This gradual evolution has obscured the full implications of the large-scale deployment of these technologies. Society's understanding of the potential environmental impacts has not kept pace with development, resulting in gaps in scientific knowledge about these impacts.

In most instances, shale gas extraction has proceeded without sufficient environmental baseline data being collected (e.g., nearby groundwater quality, critical wildlife habitat). This makes it difficult to identify and characterize environmental impacts that may be associated with or inappropriately blamed on this development.

Some of the possible environmental and health effects of shale gas development may take decades to become apparent. These include the creation of subsurface pathways between the shale horizons being fractured and fresh groundwater, gas seepage along abandoned wells, and cumulative effects on the land and communities. Similarly, monitoring strategies, data, and information on the effectiveness of mitigation measures take time to develop, acquire, and assess.

Few peer-reviewed articles on the environmental impacts of shale gas development have been published. The reasons include the young age of the industry; the proprietary nature of much industry information (in part because technologies are evolving rapidly and are still being tested); the confidentiality surrounding settlement of damage claims; and the absence of regulations for many of the chemical additives used in hydraulic fracturing (the industry therefore has not had to monitor its impact). Where peer-reviewed studies have been published, they do not necessarily agree (e.g., on the extent of fugitive methane emissions).

Information concerning the impacts of leakage of natural gas from poor cement seals on fresh groundwater resources is insufficient. The nature and rate of cement deterioration are poorly understood and there is only minimal or misleading information available in the public domain. Research is also lacking on methods for detecting and measuring leakage of GHGs to the atmosphere.

Full disclosure of chemicals and the chemical composition of flowback water is a necessary but insufficient step in the assessment of the environmental risks associated with drilling and fracturing. Information is also required on potentially hazardous chemicals produced down-hole by chemical interactions under high temperature and pressure. This includes information on concentration, mobility, persistence in groundwater and surface water, and bio-accumulation properties, for each chemical on its own and as a mixture. This represents a major gap in understanding of the potential environmental and human impacts of hydraulic fracturing, and of how to mitigate accidental releases of chemicals or flowback water to the environment.

Shale gas development also raises social impacts about which little is known. Shale gas development would take place in populated rural and semi-rural areas. Many of the people living in these areas rely on private water wells.

MONITORING APPROACHES

Reliable and timely information, including characterization, underpins the implementation of a risk management framework. Although monitoring is no substitute for effective prevention practices, it is the means by which environmental and human health impacts are identified, making it possible for mitigation measures to be designed and implemented.

Executive Summary xv

Monitoring that has been done indicates that gas leakage into aquifers and the atmosphere is frequent enough to raise concern. Given the likely future density of gas wells, shale gas development is expected to have a greater long-term impact than conventional oil and gas development.

Appropriate environmental monitoring approaches for the anticipated level of shale gas development have not yet been identified. Monitoring programs will have to be adapted to advances in technologies and to the location, scale, and pace of future development. To gain public trust, monitoring needs to engage both the people living in affected areas and independent experts. The public will have greater faith in monitoring if it can influence the design, can access the results, and can comment.

The research needed to support improved science-based decisions concerning cumulative environmental impacts has not yet begun, and is unlikely to occur without a concerted effort among industry, government, academia, and the public in each of the provinces with significant shale gas potential.

MITIGATION OPTIONS

Managing the environmental impacts of large-scale shale gas development will require not only the knowledge provided by characterizing water and ecological systems prior to development and environmental monitoring, but also a robust management framework.

There can be advantages in "go-slow" approaches to allow for additional data collection, to permit adaptation to the implications of new information, and to encourage integration of multidisciplinary expertise. However, there may also be some negative impacts of development that cannot be eliminated, and the scientific basis for identifying areas that are particularly vulnerable has not been established.

The shale gas industry has made considerable progress over the past decade in reducing water use by recycling, reducing land disruption by concentrating more wells at each drilling site, reducing the volumes of the toxic chemicals it uses, and reducing methane emissions during well completions. Other impacts, however, such as cumulative effects on land, fugitive GHG emissions, and groundwater contamination, are more problematic. This is the case because available mitigation technologies are untested and may not be sufficient; scientific understanding is incomplete; and the design of an adequate regulatory framework is hampered by limited information. Shale gas development poses particular challenges for governance because the benefits are mostly regional whereas adverse impacts are mostly local and cut across several layers of government.

An effective framework for managing the risks posed by shale gas development would include five distinct elements:

- *Technologies to develop and produce shale gas.* Equipment and products must be adequately designed, installed in compliance with specifications, and tested and maintained for reliability.
- *Management systems to control the risks to the environment and public health.*
  The safety management of equipment and processes associated with the development and operation of shale gas sites must be comprehensive and rigorous.
- *An effective regulatory system.* Rules to govern the development of shale gas must be based on appropriate science-driven, outcome-based regulations with strong performance monitoring, inspection, and enforcement.
- *Regional planning.* To address cumulative impacts, drilling and development plans must reflect local and regional environmental conditions, including existing land uses and environmental risks. Some areas may not be suitable for development with current technology, whereas others may require specific management measures.
- *Engagement of local citizens and stakeholders.* Public engagement is necessary not only to inform local residents of development, but also to receive their input on what values need to be protected, to reflect their concerns, and to earn their trust. Environmental data should be transparent and available to all stakeholders.

These elements would need to be supported by environmental monitoring programs to supply credible, science-based information to develop and apply regulations.

The regulatory framework governing shale gas development is evolving and remains untested. Advanced technologies and practices that now exist could be effective to minimize many impacts, but it is not clear that there are technological solutions to address all of the relevant risks, and it is difficult to judge the efficacy of current regulations because of the lack of scientific monitoring. The research needed to provide the framework for improved science-based decisions concerning cumulative environmental impacts has barely begun. Because shale gas development is at an early stage, there is still opportunity to implement management measures, including environmental surveillance, that will reduce or avoid some of the potential negative environmental impacts and permit adaptive approaches to management.

Chapter 1

INTRODUCTION

Fueled by a boom in the development of unconventional oil and gas resources, the energy supply picture is undergoing a dramatic change. Shale gas is one of these unconventional resources. Several observers have characterized shale gas as a *game changer.* An abundant new energy resource, often close to major markets, and relatively inexpensive to produce. Advocates for shale gas development argue that it can help satisfy expanding global energy needs, generate employment and support economic growth, decrease reliance on petroleum from unstable and undemocratic nations, serve as a transition fuel toward a carbon-free future, and play a role in mitigating climate change. Critics have raised concerns that it will adversely affect the environment (particularly water supplies), that leakage from wells and the distribution system may obviate some or all of the environmental benefits of gas burning as compared to oil and coal, and that the large-scale availability of cheap natural gas will hinder energy conservation and the transition to the non-carbon based fuels necessary to avoid disruptive climate change. Public protests against shale gas development have taken place in Canada, the United States, the United

Kingdom, Romania, Australia, New Zealand, and elsewhere. In 2011, the government of France banned hydraulic fracturing for shale gas. In the United Kingdom, a moratorium on development was lifted in December 2012 (Comité de l'évaluation environmentale stratégique sur le gaz de schiste, 2012; Davey, 2012; République Francaise, 2013).

What is the state of knowledge of potential environmental impacts from the exploration, extraction and development of shale gas resources, and what is the state of knowledge of associated mitigation options?

Further to this main question, the following sub-questions is posed:

- Based on existing research, what new or more significant environmental impacts may result from shale gas extraction relative to conventional gas extraction?
- What are the science and technology gaps in our understanding of these impacts and possible mitigation measures/strategies, and what research is needed to fill these gaps?
- What monitoring approaches could inform the effective understanding and mitigation of impacts, what is the current state of the art and state of practice for such monitoring, and what science and technology gaps may act as barriers to effective monitoring?

What technical practices exist to mitigate these impacts, and what are international best practices? What science underpins current policy or regulatory practices internationally?

The shale gas industry continually improved its methods and the science-based literature began to grow rapidly, particularly in 2013. Thus, the system had to adjust to rapidly evolving information. The system also drew analogies from the known impacts of other industrial activities. Assessing impacts far into the future based on minimal and continually changing evidence was daunting. The main long-term impacts of shale gas

Introduction 3

development cannot be known at this time; they will become evident after the passage of decades or longer.

Rather, in response to the questions, it presents the observations and conclusions on what is known and not known about the environmental impacts of shale gas development, the options to mitigate them, and opportunities for research to fill gaps in monitoring and understanding. It is hoped that its assessment and model framework for research and managing the impacts of shale gas development will help to support an informed and constructive debate.

1.1. KEY CONSIDERATIONS

Shale gas is natural gas (mostly methane) that is tightly locked within a low permeability sedimentary rock called shale. While the location of many shale gas resources has been known for a long time, only recently have technological advances made them accessible and their recovery economically viable. Only four countries, Canada, the United States, Mexico, and Australia, are currently applying this new technology of shale gas extraction by hydraulic fracturing in horizontal boreholes, although shale gas resources exist across all continents. Shale gas extraction requires the combination of brute force and sophisticated technology. The technologies required to free the gas begin with bending the well-shaft from the vertical to drill horizontally through the shale. Subsequently, the permeability of the rock is increased by injecting a customized mix of fluids, chemicals, and proppants (typically sand) at extremely high pressure to fracture the target rock, which is typically more than a kilometre below the surface and up to a few kilometres laterally from the wellhead.

Shale gas development requires (i) large amounts of water, chemicals, and proppants for the hydraulic fracturing; (ii) land for the well pads and ancillary facilities to develop the resource; (iii) energy to power the drill rigs, pumps, and trucks; and (iv) infrastructure to gain access to sites and deliver the gas. The wastes generated — mainly contaminated flowback water that must be treated or injected into the sub-surface and air emissions, including greenhouse gases (GHGs) — are also potentially significant. Given this, and while acknowledging the potential economic benefits, the large-scale development of shale gas can potentially have negative effects on the following:

- groundwater and surface water quality, if chemicals used for fracturing, the hydrocarbon gases released by fracturing, or contaminated formation water migrate into freshwater aquifers (see Chapter 4);
- the availability of freshwater resources for other uses (see Chapter 4);
- the climate, because of GHGs released during gas extraction (although the effects could be beneficial if shale gas displaces more carbon-intensive sources, such as coal, but not less carbon intensive sources such as hydro and nuclear) (see Chapter 5);
- local air quality, because of the air pollutants emitted as a result of the many activities required to drill and complete shale gas wells (see Chapter 5);
- landscape aesthetics, wildlife habitat, and existing land uses, including agriculture and tourism, because of the many well pads, roads, and ancillary facilities (e.g., gravel pits, supply yards, pipeline rights-of-way) needed to produce shale gas on a large scale (see Chapter 6);
- seismicity, stimulated by hydraulic fracturing and the injection of waste fluids (see Chapter 6);
- human health and safety from contaminated water or air or from the many activities associated with shale gas development (see Chapter 7); and
- community well-being (see Chapter 7).

These risks do not exist in isolation and can give rise to cumulative effects. The extent to which cumulative effects occur will depend on a variety of interacting factors. These include: the prevailing legal and regulatory environment; the risk-management systems used by gas and hydraulic fracturing companies; national and international energy and climate policies (or lack thereof); the availability of appropriate scientific information to judge proposed mitigation and remediation efforts; the presence and efficacy of scientific monitoring and regulatory enforcement; and the attitudes and responses of the affected communities. The need for effective scientific baseline studies and long-term monitoring is noted. However, studies and monitoring will be ineffective if they do not take place within a regulatory environment in which data are analyzed and the results used to inform policy, improve regulations, and ensure compliance. Similarly, advocates of shale gas extol the benefits of inexpensive gas, but as for any commodity (including conventional gas); the true cost must reflect the external costs of environmental damage.

For these reasons and others, it is believed that a holistic approach is needed to evaluate both the positive and negative impacts of shale gas development. Examination of the environmental effects of shale gas development should not be isolated from socio-economic, environmental, institutional, and cultural contexts. The character and extent of impacts will depend, at least in part, on those contexts (Sethi, 1979; Erickson, 1994; Holder, 2004; Porter & Kramer, 2006; and Bamberg & Möser, 2007).

The environmental implications of shale gas development are tied to important economic and social issues such as economic diversification, energy policy, climate change, and the deployment of renewable forms of energy. The pace and form of development — and thus its impacts — are influenced by a system of evolving and interacting factors that include international energy demands and natural gas prices, existing environmental conditions, government policies and regulations, institutional capacities, technologies and practices, and public opinion.

These factors provide the context for considering the environmental implications of shale gas development and of the research and management efforts that may be needed to monitor and mitigate them. Taking a holistic view of the evidence helps identify interactions among effects that may amplify or accelerate discrete impacts. It also aids understanding of non-linear, combined, and cumulative effects of impacts over large areas and time, which might otherwise be neglected. Such an approach would require enlarging the scope of concern from one primarily focused on physical, chemical, and biological effects of shale gas development to one that fully embraces its social, economic, and political dimensions.

The need for a holistic approach is particularly clear in the domain of GHG emissions. On the one hand, advocates of shale gas development argue that shale gas will have a positive impact on climate change because natural gas releases less carbon dioxide when burned than coal or oil. In addition, they suggest that natural gas can provide a bridge to renewables, helping to supply world energy needs while carbon-neutral energy capacity (e.g., nuclear, solar, wind, biofuels) and improved efficiency is developed (Bloomberg & Mitchell, 2012; and Richardson, 2013). On the other hand, opponents argue that shale gas development reinforces the patterns of fossil fuel dependency that drive climate change, particularly when the price of gas is relatively low, by strengthening and expanding the infrastructure that supports it, a concept that historians of technology have variously described as path dependence, technological momentum, and/or infrastructure trap (Hughes, 1983; Worster, 1992; and Jones, 2010).

Both advocates and opponents of shale gas development recognize the potential for a large number of wells to be drilled in coming years. Ensuring that wells will not develop gas leaks over time is a long-standing engineering challenge that adds to the uncertainty. Analyses of shale gas GHG emissions depend to a significant extent on the amount and degree of methane leakage from active or abandoned wells (see Chapter 4). There is much uncertainty concerning the accuracy of the methods used to produce the sparse methane-leakage data that currently exist. At the low end, scientific estimates support the conclusion that GHG emissions from shale gas are comparable to conventional gas and therefore are clearly lower than coal (e.g., Cathles et al., 2012; Logan et al., 2012; and O'Sullivan & Paltsev, 2012). At the high end, estimates support the conclusion that the GHG benefits of natural gas over coal are obviated by the leakage problem (e.g., Wigley, 2011; and Petron et al., 2012).

Disagreement also prevails over what exactly is *unconventional* about unconventional gas. Advocates argue that years of experience from developing and operating the approximately 150,000 horizontal, multi-

Introduction

staged, hydraulically fractured oil and gas wells drilled in to date demonstrate substantial economic benefits and reveal no fundamental environmental harms (e.g., Zeirman, 2013). Opponents argue that shale gas development has disrupted the quality of life in rural regions, contaminated groundwater and surface water resources, damaged sensitive habitat, and undercut the market for renewable energy (e.g., Mufson, 2012; and Wilson et al., 2013). Some of these claims, particularly those related to the migration of hydraulic fracturing fluids from deep underground into regional groundwater resources, are difficult to evaluate because of a lack of baseline data and scientific monitoring, and because the time-frame in which adverse effects may manifest is long. Claims there are *no proven adverse effects* on groundwater from shale gas development lack credibility for the obvious reason that absence of evidence is not evidence of absence. Further, groundwater has been affected due to incidents such as loss of containment due to faulty well casings or leakage from holding ponds.

The strongly contrasting views of shale gas development point to the need for much more extensive and comprehensive studies. They also point to the need to consider past experience when dealing with new forms of environmental risk. Retrospective analysis suggests that western societies — driven by technological optimism and a belief in the desirability, if not inevitability, of economic expansion — have often underestimated the risks posed by the introduction of new technologies. The detailed study by the European Environment Agency, *Late Lessons from Early Warnings*, documents numerous examples in which evidence of adverse environmental impacts from economic activity was discounted based on justifications that seemed logical at the time but turned out to be incomplete at best. These examples include factors that are relevant here, such as the demand for employment and economic development, and the tendency of advocates for new technologies and economic activity to assert that a lack of proof of harm is equivalent to a proof of safety. At this stage in shale gas development, there are many unanswered questions. This should be taken as an indication of the need for more and better information, as well as the need to ensure that existing information is made available to relevant stakeholders and used in a fair and intelligent manner.

Shale gas development is already significantly affecting the economy and government revenues and has the potential to do so in at least four other provinces. In regions near towns or cities, shale gas development will compete for resources (e.g., water), offer new employment, and place significant demands on existing infrastructure. It may also represent the start of several decades of production, the drilling of tens of thousands of wells, and possible long-term impacts after well closure and abandonment.

This development will have both local and regional effects; evaluation of the potential environmental impacts cannot, therefore, focus on a single well or well pad, but must also consider regional and cumulative effects.

1.1.1. Regional Differences

The shale gas regions range from near urban and populated rural areas in the south to wilderness in the northwest. The expression *environmental impacts from shale gas development* masks many regional differences that are essential to understanding these impacts. The main regional differences include:

- population density and related use of local water resources;
- the chemical composition of the natural gas: some is dry (without significant secondary gases), some contains natural gas liquids, and some has high carbon dioxide content;
- the regional geology (i.e., the depth, thickness, and composition of the shale rock and the presence of natural fractures or faults and natural stress fields);
- the surface features (e.g., hydrology, remote or settled area, boreal forest or agricultural land, dry or wet climate);
- the technology used for fracturing (e.g., whether it uses water, how much water, what kinds of chemical additives);
- the regulatory regime setting the rules by which activities are permitted;
- the extent of competition for potentially scarce resources, including agricultural land, surface water and groundwater supplies, and unspoiled rural conditions;

Introduction

9

- the existence of Native rights and title issues; and
- the overall social context in which the development takes place.

One of the major regional differences in the societal aspect of shale gas development is that much of the land with potential shale gas reserves is privately owned. However, the land is owned outright by the provincial government and the sparse population is primarily there has been ongoing oil and gas development for many decades with public acceptance. By contrast, although petroleum resource development has historically taken place in some areas, no significant activity of this type has occurred in the past half century in the shale gas regions. Therefore, shale gas development will be new for today's residents in these areas.

The assessment of these differences requires detailed consideration of the characteristics of each region. Consequently, the discussion in this volume about the potential environmental impacts from the development of shale gas resources is necessarily general in nature since large-scale domestic shale gas production is only taking place are still several years away from producing shale gas on a large scale, if at all. In addition, the potential environmental impacts associated with shale gas development are diverse: impacts in one region may not occur in another; practices that are appropriate in one region may be unacceptable elsewhere.

1.1.2. Public Trust

All development takes place in a social context. Which environmental impacts are considered acceptable will vary by region, depending on factors such as prior familiarity with the gas industry, the nature of potential land use conflicts, the socio-economic context, the level of trust in industry and government, and the industry's performance.

Although one can debate the extent to which shale gas development involves new technologies rather than the natural extension of existing technologies, the industry has touted high volume hydraulic fracturing and horizontal drilling for shale gas as new technologies and the public perceives them as such. This raises the possibility of new (and possibly unknown) environmental risks. The public seeks an assurance that these risks either do not exist or are small enough that they can be managed satisfactorily. To meet this demand, the industry and government regulators will have to monitor and document environmental impacts in a way that is transparent and credible and earns public trust. Public acceptance will not come based only on industry claims of technological prowess or government assurances that environmental effects are acceptable.

These observations apply equally well in the context of shale gas development. People's perceptions of the environmental risks associated with large-scale shale gas development need to be considered along with the risks themselves. A discussion of the factors that impact public perceptions of chemical risks (Box 1.1).

Box 1.1. A Summary of Key Factors that Affect Public Perceptions of Acceptability for Chemical Risks

- The distribution of risks and benefits is more important than the balance of risks and benefits.
- Unfamiliar risks are less acceptable than those considered to be familiar.
- Hazards that invoke dread are perceived more negatively, even when the risk level is low.
- Risk that is voluntarily taken is more acceptable than a risk that is imposed.
- Risks that people feel they can control are more acceptable than those they cannot.
- Risks imposed by unethical actions are perceived negatively.
- Anthropogenic risk is generally less tolerable than "natural" risk.
- Relative risk is more significant than absolute risk.
- Trust in the risk manager is critical.

(Covello, 1983, 1992)

Introduction 11

1.1.3. The Evidence

The state of the evidence available for this literature is in a variety of forms distributed across many disciplines and is of varying quality: peer-reviewed scientific, engineering, and social science articles, government reports, consultants' reports, industry studies, non-governmental organization studies, and mass media coverage. Some of it comes to conclusions on the same or comparable questions. Much of this literature is from, reflecting the longer history of shale gas development, and many studies are ongoing. While the geological environments, local conditions, and regulatory environments in these studies are nevertheless important to consider. An important difference between shale gas developments in the pertaining to public opinion is local economic benefits. In many parts of the world, mineral and petroleum belong to the owner of the land, leading to an incentive for development as these owners gain substantial financial benefits from payments made by shale gas companies. However, provincial governments own all petroleum and mineral rights and the payments to land owners are for access to drill wells.

Even as the literature on the environmental effects of shale gas development is growing, there are several challenges faced concerning the state of the relevant evidence:

- In nearly all instances, shale gas extraction has proceeded without important environmental baseline data being collected (e.g., nearby groundwater quality). This makes it difficult to identify and characterize environmental impacts that may be associated with (or incorrectly blamed on) this development.
- There is a paucity of peer-reviewed articles in the scientific literature. The reasons include the fact that large-scale shale development is a young industry, that the industry has kept some information proprietary (in part because technologies are evolving rapidly and are still being tested), the chemical additives used in hydraulic fracturing and therefore industry has not had to monitor their impact.

- A major environmental concern regarding shale gas development — regional groundwater contamination — hinges on the flow of fluids in low permeability but commonly fractured geological strata. However, because past scientific interest has largely focused on high permeability rocks (aquifers and petroleum reservoirs); fluid flow in low permeability rocks is poorly understood. Thus, the basic scientific knowledge needed to evaluate potential risks to groundwater on the regional scale is largely lacking.
- In areas where peer-reviewed studies are available, they do not necessarily agree. For example, there is a substantial range of expert opinion on the extent of fugitive methane emissions from shale gas development.
- Some of the possible environmental effects of shale gas development, such as the creation of sub-surface pathways between the shale horizons being fractured and fresh groundwater, gas seepage from abandoned wells, and cumulative effects on the land and communities, may take decades to become apparent. Similarly, monitoring information, and information on the effectiveness of mitigation measures, take time to acquire and assess.
- Much if not most of what can be said about the potential environmental impacts of shale gas development depends on assumptions made about the location, pace, and scale of development, all of which will be influenced by future natural gas prices, government policy, and technological improvements. None of these can be predicted with certainty.
- Given these challenges, the literature features a range of diverse views on the environmental effects of shale gas development. Experts from different scientific disciplines disagree on the risks posed by hydraulic fracturing, and people living close to shale gas development have their own views about these risks. These disagreements explain why many conclusions to qualified. They also explain why many areas of incomplete scientific knowledge identified and understanding of the environmental effects of shale gas development.

Introduction 13

1.2. MANAGING IMPACTS

It is essential to identify practices and approaches for monitoring and mitigating impacts, and to identify information deficits that may act as barriers to effective monitoring. These are crucial questions that do not lend themselves to purely technical answers. Managing the environmental impacts of large-scale shale gas development will require:

- the application of sound technologies,
- rigorous management systems,
- appropriate outcome-based regulations with strong performance monitoring, inspection, and enforcement,
- the recognition of regional differences, and
- effective public engagement.

This section elaborates on these five elements of a management approach in Chapter 9.

Some of the adverse impacts of shale gas development can be avoided by applying existing good practices. Others can be managed and kept to acceptable levels, although the threshold of public acceptability will vary regionally. Some risks, such as cumulative impacts on the land and contamination of groundwater, are more problematic: either we do not know enough about the probability of the risks or, where we do, they may force difficult trade-offs. As outlines in Chapter 9, the rules for managing the environmental effects of shale gas development are becoming increasingly sophisticated, and it is important to ensure that they are applied.

However well-crafted, rules will not suffice if they are not supported by good-quality environmental information and enforcement. As explained above, much of the information required to assess the environmental risks posed by shale gas development either does not currently exist or is not publicly available. In fact, given the nature of the extraction and monitoring technology, some of the information will not exist until the technologies are applied, and/or until government regulators or other authorities insist that it is collected and analyzed.

This underscores the importance of environmental monitoring based on science, both to understand risks and impacts, and to design and evaluate the effectiveness of mitigation measures. Through monitoring, predictions about impacts can be tested, measures can be put in place or adapted to reduce or eliminate risks, and public concerns can be addressed. Any monitoring program will have to be adaptive to the research needed to determine how to monitor effectively.

Monitoring must also engage the people living in the areas that will be affected by development: they must be able to influence what is monitored, have access to the results and be able to comment on them, and be able to participate in decisions about appropriate responses.

One obvious difference between shale gas and conventional gas development mentioned above is the scale of development, which will lead to much greater local impacts, including potential community disruption, where there are substantial shale gas resources, shale gas development would make unavoidable the drilling of gas wells relatively close to water wells relied on for drinking by rural residents. This scale of development may have substantial social, economic, and cultural ramifications, as it may affect quality of life and community and individual well-being.

A second important difference is that large quantities of liquid are pumped under extremely high pressure into deep formations to extract the hydrocarbon resource. About one-quarter to half of the water used in hydraulic fracturing returns up the well to the surface. This flowback typically is hazardous, containing a portion of the hydraulic fracturing chemicals, hydrocarbons (including variable amounts of benzene and other aromatics), and constituents leached from the shale such as salt, metals, metalloids, and natural radioactive constituents. Although flowback water is frequently reused in fracturing operations, a proportion often remains, posing technical problems for treatment and disposal in some regions.

A third difference involves gas leakage from wells. Even when oil and gas wells are sealed with cement using best industry practices, some will leak because of difficulties in establishing a continuous cement seal along the steel well casing, because of gradual deterioration of the cement seal over time, or because of damage to the rock formation adjacent to the casing creating open fractures. This leakage may allow gas to leak from the well into the atmosphere, and gas and other fluids to seep from along the well into adjacent geological strata and shallow groundwater and/or create a short circuit between otherwise hydrogeologically isolated geological horizons. There are found little literature assessing the impacts of gas leakage in the conventional oil and gas industry. However, the potential number of wells in shale gas development is very large, increasing the potential impact of gas leakage. In addition, the potential number of hydraulically fractured wells in areas where rural and suburban populations rely on groundwater wells is much larger than for the typical areas of conventional oil and gas development in the past half century. As a result, the greater effects of this type of well leakage in the shale gas industry are an important issue to be considered.

Uncertainty is inherent with any large-scale technological development. In the case of shale gas, this uncertainty is magnified because of the limited quantity and quality of the relevant scientific information. Management approaches will therefore need to foster the development of appropriate information, adapt to new knowledge as it is acquired, make informed decisions as to which projects should proceed and which should not, and implement new mitigation measures or modify existing ones during the life of projects that do proceed.

1.3. HOW THIS VOLUME IS ORGANIZED

Rather than aligning the book with the individual questions posed chose to organize the book into three main sections:

- Background and context for the book are provided in Chapters 2 and 3;
- Environmental and health impacts of shale gas development, are explained in Chapters 4, 5, 6 and 7; and
- Managing and monitoring impacts are discussed in Chapter 8 and 9.

The following points summarize the relationship between the sub-questions and the structure of the book:

- Based on existing research, what new or more significant environmental impacts may result from shale gas extraction; Addressed in Chapters 4, 5, 6 and 7.
- What are the science and technology gaps in our understanding of these impacts and possible mitigation measures/strategies, and what research is needed to fill these gaps? Addressed in Chapters 4 to 9.
- What monitoring approaches could inform the effective understanding and mitigation of impacts, what is the current state of the art and state of practice for such monitoring, and what science and technology gaps may act as barriers to effective monitoring? Addressed in Chapters 8 and 9.
- What technical practices exist to mitigate these impacts, and what are international best practices? What science underpins current policy or regulatory practices internationally? Addressed in Chapter 9.

Chapter 2

SHALE GAS DEVELOPMENT IN THE CONTEXT: WHAT IS SHALE GAS?

Shale gas is a natural gas composed primarily of methane (CH_4; more than 90 percent) found in organic-rich shale formations. Shale itself is a sedimentary rock made up predominantly of consolidated clay- and silt-sized particles. Shales are deposited as mud in the quiet waters of tidal flats, deep-water basins, and similar low-energy depositional environments. Algae-, plant-, and animal-derived organic debris becomes mixed in when these very fine-grained sediments are deposited.

As mud turns into shale over geological time, bacteria metabolize the available organic matter and release biogenic methane as a by-product. Because it remains shallow — generally just a few hundred meters deep — this biogenic methane will sometimes naturally seep into groundwater and may even infiltrate water wells. Aquifers in many sedimentary basins worldwide contain some dissolved methane (Barker & Fritz, 1981; and AESRD, 2011). Natural gas also forms during deep burial, when the organic matter is cracked from high pressure and heat, converting it into lighter hydrocarbons, creating thermogenic methane.

Just like conventional gas, most of the shale gas produced is thermogenic methane. It may contain small amounts of other gases (e.g., ethane, butane, pentane, nitrogen [N_2], helium, carbon dioxide [CO_2]) and impurities, as does conventional gas. Like conventional gas, shale gas can also be *wet* (contains commercial amounts of *natural gas liquids* like ethane and butane) or *dry* (contains very little or no natural gas liquids). Thermogenic methane can be differentiated from biogenic methane through isotopic analysis (see Chapter 4).

Some of the hydrocarbons produced in shale gas reservoirs manage to escape and migrate into the more permeable rock — typically sandstones — and remain trapped underground by a seal of very low-permeability rock. These are referred to as conventional reservoirs. The majority of the world's conventional hydrocarbon reserves were generated from organic-rich shales, from which they then escaped. Oil and gas in conventional reservoirs are fairly mobile, able to rise easily until they become trapped against a ceiling of rock of low permeability. However, shale gas, both biogenic and thermogenic, can be found in three forms where it was first generated:

- as free gas in pore spaces/fractures;
- as adsorbed gas (i.e., where gas is electrically stuck to organic matter/clay); and
- as gas dissolved in organic matter (only a small amount is in this form)

As a result of being in such very low-permeability reservoirs — the pores in a shale formation can be 1,000 times smaller than those in conventional sandstone reservoirs — shale gas is considered to be *unconventional*, requiring special completion, stimulation, and/or production techniques to be economically produced. Note that shale is not homogeneous and can vary greatly in mineralogical composition, geochemistry, and geomechanical behaviour, even over short distances. Thus, every type of shale is different and the well completion technology used to extract the gas must adapt to these variations (e.g., Halliburton, 2008). It is also worth noting that reservoirs made of tight sandstones are of low permeability and are therefore considered unconventional, requiring fracturing to release and mobilize the gas.

2.1. BACKGROUND

Natural gas plays an important role in the economy, meeting energy needs and representing a large source of export revenues. It is used extensively in residential, commercial, and industrial markets and, to a lesser extent, for power generation. Natural gas burns more cleanly than do other fossil fuels, emitting fewer air pollutants and less carbon dioxide (about half that of coal), thus contributing less per unit of energy to the GHG emissions.[1]

2.1.1. Social Context

The large-scale commercial deep shale gas industry is still young. Pushed by rapid technological innovation and rising gas prices, its growth has been spectacular and given rise to much public concern, particularly in areas with little or no previous gas industries (e.g., eastern North America and Europe). These areas also happen to be more densely populated and have more intensive land uses, thus presenting more difficult trade-offs than many traditional conventional gas-producing areas.

Several incidents of water contamination, cattle deaths, induced earthquakes, and regulatory violations during the rush to development in the industry's early days have fuelled concerns and led to opposition, public protests, and civil suits (Committee of Energy and Commerce, 2011; ALL Consulting, 2012; Infante et al., 2012; Pembina Institute, 2012; and Royte, 2012). Public concern has further deepened with the secrecy of several operators, about the chemicals used in fracturing fluids, for example; by the nuisances and disruptions associated with industry activities; and by industry playing down the environmental risks involved.

Although the nature of the concerns expressed varies regionally, certain themes concerns are widely shared:

---

[1] *Marketable reserves* refers to gas that can be produced economically at current prices and with known technology.

- Methane and fracturing fluid contaminating drinking water. Incidents involving methane contamination of tap water have received widespread media coverage.
- Air emissions (including GHGs) harmful to public health and contributing to climate change.
- Seismicity induced by hydraulic fracturing or deep-well waste injection. These concerns led to an independent review of shale gas development in the United Kingdom by the Royal Society and Royal Academy of Engineering.
- The scale and pace of development. Critics argue that jurisdictions with little or no previous oil and gas development are unprepared to regulate the high level of activity projected for shale gas.

In populated areas, residents have expressed concerns about:

- the impact of development on property values;
- the effect of possible accidents (e.g., blow-outs, spills, operational malfunctions); and
- the diminished quality of life associated with shale gas operations (e.g., noise, dust, smells, strong lighting, truck traffic).

In rural areas, concerns include:

- contamination of domestic farm well water;
- conflicts with existing land uses including the loss of agricultural land to shale gas development and ancillary activities (well pads, quarries, roads, pipelines, staging areas, etc.);
- damage to rural roads by heavy truck traffic;
- changing social conditions as a result of the influx of outside workers;
- the amount of fresh water the shale gas industry might use;
- potential adverse effects on aquaculture stemming from a loss in water quality; and

Shale Gas Development in the Context: What Is Shale Gas?   21

- loss of subsistence hunting, trapping, and berry-picking opportunities.

(Lapierre, 2012)

**Table 2.1. Summary of Local Concerns Expressed
About Shale Gas Development**

Selection of comments from elders, land users, and other knowledge holders	• Access roads disrupt wildlife • Rivers and lakes contaminated • Dust and noise from increased road traffic • Surface water withdrawals affect water levels • Chemical spraying kills berries and affects plants and wildlife • Site inadequately remediated
Based on public hearings	• Garbage attracts bears • Excessive hunting by non-First Nations • Development too close to villages and water sources
Based on comments consultations	• Water contamination • Protection of water supplies • Waste management
Based on comments received	• Air quality • Greenhouse gas emissions • Technical and seismic risks
Other surveys	• Traffic • Noise • Light pollution • Local air quality • Groundwater contamination • Spills • General distribution • Property trespass and damage

Table 2.1 summarizes the environmental concerns voiced in different Shale Gas Development. Many concerns relate to the cumulative impact local residents think shale gas development will have on their quality of life. Although a few wells might be acceptable, many see the prospect of large-

scale development as a threat to the values they cherish — a clean environment, rural tranquility, access to the land, and acquaintance with all their neighbours. In addition, many individuals and groups have expressed their concern about what they perceive to be inadequate government oversight and readiness. They do not believe that their governments have the capacity to regulate the industry effectively and protect the environment while maximizing economic opportunities (BAPE, 2011b; Lapierre, 2012).

Note that, while public opposition to shale gas development has been substantial in many individuals, business associations, and other organizations have spoken in favour of the economic benefits they believe such development can bring (e.g., jobs, government revenues, regional development, new energy supplies) (BAPE, 2011b; Lapierre, 2012).

CONCLUSION

Geological estimates indicate that holds a vast shale gas potential, which dwarfs remaining conventional natural gas reserves. While shale gas is already being produced in in increasing volumes, most of this potential remains unexplored. Early drilling results imply that shale gas could one day be produced in significant volumes from regions with little tradition of petroleum development, such as. The economic, social, and environmental implications are likely to be far-reaching.

The prospect of shale gas development has given rise to numerous public concerns in the regions affected, most related to the possible environmental implications of such development. These concerns, as well as the sharp drop in natural gas prices in the past few years, have considerably slowed down the pace of development. All affected provincial governments have responded to the prospect of shale gas development and expressed public concerns by updating their relevant policies and regulations and launching various studies on the possible impacts of shale gas development.

Chapter 3

SHALE GAS TECHNOLOGY AND WELL INTEGRITY

To understand the environmental implications of shale gas development, it is necessary to understand the technology used and how it is evolving. This chapter describes the main steps in shale gas development, from initial exploration, to the construction of well pads and associated infrastructure, to drilling and well completion (including cementing, hydraulic fracturing and the use of chemical additives).[2] The emphasis of this discussion is on the importance of well integrity[3] and the importance of preventing gas leakage concluded that this aspect of the technology is of paramount importance in long-term environmental protection.

Shale gas has been produced for decades from geological formations with natural fractures that allow economical recovery from shallow vertical wells producing at low rates over a long time. Improvements in technology and increases in gas prices enabling the large-scale commercial production of much deeper shale gas reservoirs.

---

[2] See NYSDEC, 2011; ALL Consulting, 2012; King, 2012 for a more detailed discussion of these steps.

[3] Well integrity refers not to the ability of the steel casing to maintain internal pressure, but to the capability of the well to prevent leakage of gas and other fluids upward into the Fresh Groundwater Zone and the atmosphere.

Horizontal drilling is generally reserved for deep wells, usually more than one kilometre, because it is cheaper to drill a larger number of vertical wells at shallow depths. In addition, shallow horizontal wells pose a greater environmental risk.

Multi-stage hydraulic fracturing involves injecting a fluid — usually water, but sometimes gas or a petroleum-based liquid, plus chemicals and proppants (generally sand) to improve fracture placement, performance, and gas recovery.

Well integrity refers not to the ability of the steel casing to maintain internal pressure, but to the capability of the well to prevent leakage of gas and other fluids upward into the Fresh Groundwater Zone and the atmosphere at extremely high pressure into the shale formation in a number of places (stages) along the wellbore. This is done to fracture (hence the term *fracking*) the rock and create a network of open fractures through which the gas can flow. The fracture network is made up of existing fractures, formerly closed fractures, and new fractures.

What is new is the combination of these two technologies; the use of greater amounts of water, sand, and chemicals; and the higher injection rates and pressures to fracture a much larger volume of rock. What is also new is that these technologies are now being applied much more widely to a broad spectrum of unconventional oil and gas resources. Apart from shale gas deposits, shale oil, tight oil, and tight gas strata are also being stimulated through high-pressure, multi-stage hydraulic fracturing. It is also important to note that the application of these technologies continues to evolve rapidly. For example, horizontal laterals have grown longer; the composition of additives in fracturing fluids has been modified with experience; more stages are being done per well at higher injection rates), reducing the overall well pad density. However, physical limits and cost optimization considerations indicate that one should not extrapolate trends from these examples.

Apart from hydraulic fracturing, the scale of the development is what differentiates shale gas development from conventional gas development. Although both conventional and shale gas development require the construction of well pads, work camps, roads, and pipelines, shale gas development requires more of these activities (as well as hydraulic fracturing) because:

- the reach of individual wells in low-permeability rock is far less than it is in highly permeable rock; and
- the production of individual wells declines faster so more wells are needed to sustain a stable production rate.

Figure 3.1. Surface Infrastructure of a Hydraulic Fracturing Treatment Site. The equipment at a shale gas site during hydraulic fracturing. The inset numbers correspond to the following activities: (1) Data / Satellite Van: To monitor and control treatment operations. Satellite transmission of data to enable real-time monitoring from remote locations, (2) Sand storage silos, sand conveyors, two day silos over blenders: To store proppant (sand) and to feed it to the blender mixing tub (underneath), (3) Multiple Wellheads (shrouded for safety reasons during the stimulation operation), (4) Pumping units: High-pressure pumps to pump the fluid, (5) Chemical vans and tanks: Storage, transportation, and metering units used to feed additives into fracturing fluid stream as it is pumped down the well, (6) Test equipment: To receive and measure flowback water in a controlled manner. The equipment is also used to separate fluids from gas, which is sent be flared, (7) Water surge tanks: Tanks to prevent pressure surges and provide smooth water supply to the operation. In this operation fresh water was pumped from storage ponds. No flowback water is stored on the surface, it is pumped directly to disposal wells after separation from the produced gas. In some jurisdictions, lined ponds may also be used to store flowback water that is treated and reused. The two cranes on site are suspending microseismic sensor arrays in two of the wells and

the nearby trucks are recording the microseismic data (Reproduced from Nexen Energy ULC website).

Thus, even with multi-well pads, shale gas development will lead to more pads being built and many more wells being drilled than would be needed to produce the same volume from conventional gas reserves in high permeability reservoirs.

Figure 3.1 shows a photo of the surface infrastructure at a shale gas site during hydraulic fracture stimulation with a description of the main equipment being used.

Shale gas wells with longer horizontal laterals typically take longer and cost more to complete than conventional vertical wells to the same depth. As with any other commodity, shale gas production is sensitive to price. The rise in gas prices early in the 21st century per thousand cubic feet was an important driver of technical innovation that increased shale gas production. In contrast, the recent recession as well as mild winter weather and glut in supply have driven down the price of natural. Until prices recover, the economics of expanding shale gas production, particularly dry gas (without synergistic production of larger organic molecules such as ethane and heptane), remain in question.[4] In the following sections of this chapter, each stage in shale gas development is described with an emphasis on its potential environmental impacts.

3.1. PREPARATION FOR THE WELL DRILLING STAGE

The first stage in shale gas development, *exploration*, involves scanning the subsurface geology using geophysical methods, mainly ground-based seismic surveys. Next, a few wells are drilled and rock cores collected, mostly from the geological strata where shale gas resources are known or expected. Geophysical measurement tools are run down these holes to obtain additional insights about the geology, porosity, permeability, and other properties of the subsurface. Unless they are to be used later as

---

[4] Because ethane, propane, butane, and natural gas liquids (*pentanes plus*) command higher prices, industry is currently focusing on developing wet shale gas reserves.

monitoring wells (usually for microseismic monitoring) these exploration wells are sealed with cement at the surface after the geophysical logging is completed.

In shale gas plays, this exploration phase is relatively short as the formation to be produced is generally well known. This applies to most parts where substantial information on the geology of all the shale gas reserves is available. This exploration stage therefore focuses on identifying the most favourable areas for development.

Well pads must be constructed before production wells can be drilled. Such construction also occurs in conventional oil and gas development. For shale gas development, the pad needs to be larger to accommodate the drilling of multiple wells and the large amount of equipment, chemicals, and sand used in the multi-stage hydraulic fracturing operation. In conventional oil and gas development, the pads are typically 0.5 to 1.0 hectares, whereas in shale gas development they are generally 2.0 to 3.0 hectares. The pads must be nearly horizontal and built with good quality fill approximately 0.5 to 1.5 meters thick, depending on the nature of the subgrade.

After removing vegetation from the site, the organic-rich soil at the pad is scraped off and stored for partial site reclamation after all wells are completed. The well pad itself is typically constructed with fill excavated at or near the pad, and the resulting borrow pit is often used to store the water to be used in fracturing. If no local fill is suitable, access to a more distant source is necessary, with trucks hauling the granular pad construction material. A large pad for 16 shale gas wells may take up to several months to construct and up to 500 to 800 truckloads of fill if none is locally available and if the subgrade is so weak that a load-bearing surface pad is required.

In northern climates, pads are constructed in the summer, when fill compaction is easier and wells completed in the winter, when the ground is frozen and can withstand the heavy loads of road and pad traffic.

Before carrying out a multi-stage hydraulic fracturing operation in a new area, oil and gas companies undertake geoscience and geomechanics studies in part to estimate the maximum height growth of induced fractures so that they will not extend significantly above the shale gas prospective horizon. Such estimates are based on mathematical modelling and on previous experience in similar conditions, and are refined by measurements during and after the hydraulic fracturing stimulation operations.

Where uncertainty remains as to the details of hydraulic fracture propagation, companies use microseismic methods to monitor the first multi-stage hydraulic fracturing treatments, sometimes in several wells. In the early stages of a field development, continuous monitoring yields much information about the extent and height of the stimulated zone. After a relatively small number of wells in a region are hydraulically fractured, enough knowledge has generally been gained to make microseismic monitoring for subsequent wells in the region unnecessary (i.e., the empirical response model is reasonably well calibrated).

This empirical approach to the design of multi-stage hydraulic fracturing treatments, based on data, is necessary because the mathematical models for fracture propagation prediction in naturally fractured rock are weak and must be calibrated (Dusseault, 2013). Even with the calibrations, considerable uncertainty remains, but not to the extent that fractures could propagate in an uncontrolled manner. They indicate that public confidence would be increased if microseismic monitoring was done more often.

Once full-scale drilling is under way in the specific area, no additional microseismic monitoring is done and the improvement in the understanding of the system and productivity evolves from the empirical response model. This model is an optimization approach that uses real performance data to seek the best multi-stage hydraulic fracturing approach (volume, rate spacing, fluids, etc.) for a well with a set of physical parameters (thickness, permeability, natural fracture intensity, stiffness, *in situ* stress fields, etc.), so that future wells can be developed close to their maximum potential, given typical levels of uncertainty in the reservoir parameters.

3.2. WELL DRILLING AND COMPLETION

Once the well pad is constructed, well drilling can begin. At different stages during drilling, casing strings (attached joints of steel pipe) are run into the hole and cemented into place. Operators try to achieve well integrity by installing barriers between the fluids used for drilling and those produced from the reservoir and the environment, such as multiple casing strings, a cement sheath around the casing, and a blow-out preventer (BOP).

The first drilling, usually carried out by a small auger rig, installs a conductor pipe about 10 to 15 metres deep that will allow the drilling fluid to be properly returned to the surface and continuously reconditioned. The surface conductor pipe is about 70 millimetres in diameter, and is roughly cemented into place by pouring cement around the outside of the shallow pipe.

A large drill rig can be moved to the location, and the surface casing hole is drilled using a non-hazardous bentonite-water slurry. The surface casing is designed to protect surface water and shallow aquifers from cross-flow and contamination and ideally is placed below the lowest zone of potable groundwater (about 200 to 300 metres) (Figure 3.2). The surface casing is cemented into the borehole to achieve these goals and to provide structural support for the future wellhead and the BOPs that are installed before the next drilling phase. The casing is cemented by pumping cement down it and up the annulus (the area between the casing and the borehole wall), and the casing is equipped with centralizing and scratching devices to remove mudcake and place the casing in a concentric position. In addition, various casing rotating and reciprocating actions are taken to ensure a continuous cement sheath with no gaps.

Once the surface casing is installed, the hole is deepened. An intermediate casing string is often installed, perhaps to a depth just above the shale to be produced, typically 1,500 to 3,500 meters, to keep the borehole stable. This intermediate string must be cemented to the surface.

It is intended also to isolate any non-commercial gas and oil zones it may intersect. In fact, proper isolation in this intermediate depth region may be the most important factor in preventing contamination of fresh groundwater resources and preventing gas escaping to the atmosphere (see Chapter 4).

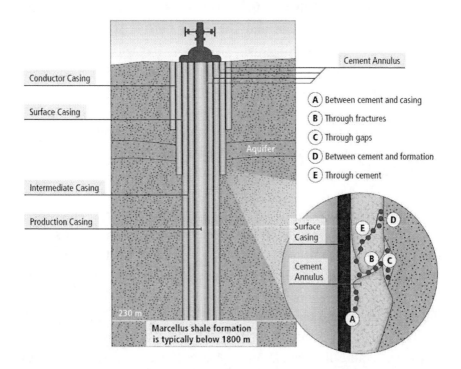

Figure 3.2. Typical Marcellus Well Construction. Figure illustrating a well (i) The conductor casing string forms the outermost barrier closest to the surface to keep the upper portion of the well from collapsing and it typically extends less than 12 metres from the surface; (ii) the surface casing and the cement sheath surrounding it that extend to a minimum of 15 metres below the lowest freshwater zone is the first layer of defense in protecting aquifers; (iii) the annulus between the intermediate casing and the surface casing is filled with cement or a brine solution; and (iv) the production string extends down to the production zone (900 to 2,800 metres), and cement is also placed in the annulus between the intermediate and production casing. Potential flaws in the cement annulus (inset, "A" to "E") represent possible pathways for gas migration from upper gas-bearing formations or from the target formation (Data Source: Vidic *et al.*, 2013).

The well is further deepened and, as it approaches the shale gas horizon at a depth of 1.5 to 4.0 kilometers, it is turned to the horizontal. Depending on what technology will be used for multi-stage hydraulic fracturing, the drillhole may be extended to the full length of the horizontal section, or drilling may stop at the point where the hole becomes horizontal. In either case, the production casing is then installed and cemented to the surface or at least well into the intermediate casing to achieve proper zonal isolation.

For some multi-stage hydraulic fracturing technologies, the horizontal section is drilled from the shoe that was placed just after the well became horizontal, and an open-hole completion is carried out using highly specialized equipment designed to facilitate the staged fracturing operation to come. At this point, the well is ready for multi-stage hydraulic fracturing. On a multi-well pad, the drill rig is skidded in the vertical position to the next conductor pipe and the process repeated until all wells are drilled, completed, and equipped for multi-stage hydraulic fracturing.

A typical simple wellbore may have, for example, the upper 250 meters drilled at a diameter of about 13 inches, with installation of a 10½ inch outer diameter casing. Drilling continues to the full depth with a 9-inch bit, and a 7-inch production casing is installed. Casing strings are tested for pressure integrity at each stage of the drilling operation so that remedial action can be taken in the event of a casing breach or a leaky connection before proceeding to the next stage. Testing involves closing the annulus and pressurizing it with water to verify that the pressure can be held without leakage. The cementation quality is also assessed by running a cement bond log device down the well. The log measures the response to a continuous acoustic signal transmitted through the steel casing and beyond. A pipe that has no cement, significant gaps, highly irregular cement sheath thickness, or a weak steel-cement-rock bond will reflect different and larger amplitude acoustic signals than pipe and rock that are strongly bonded concentrically with good quality cement. However, these logs are not capable of identifying all the potential pathways behind the casing (such as microcracks in the rock) that could be involved in well leakage.

If severe problems are identified, such as a zone where drilling fluid was not completely displaced during cementation, the installed casing may be perforated, the cement or resin squeezed out at high pressure, and a thin steel liner placed across that zone to help isolate it.

If water-based fluids are used during drilling, the separated well drill cuttings (the bits of rock ground in the drilling process) are almost always environmentally benign and may be buried near the site. If oil-based fluids are used, the cuttings must be treated according to provincial regulations to prevent surface contamination. Bentonite-based drilling fluids are generally used for the next well on the pad.

Because organic matter tends to adsorb thorium and uranium ions that may be moving through the deep-water flux system, shale gas zones often have natural background radioactivity higher than other strata. Nevertheless, these Deep Zone cuttings are still well below the threshold values that would require special isolation techniques for the drill cuttings disposal.

3.2.1. Hydraulic Fracturing

Fracturing is usually performed by specialized service companies rather than the oil and gas company operating the well. Perhaps large- to medium-sized service companies have the right equipment and sufficient experience to execute large multi-stage hydraulic fracturing operations on multiple wells on a large pad, but dozens of smaller companies may be involved in the process. These smaller companies provide specialized services such as perforating, materials provision, pumping capacity, bottom-hole assembly fabrication, data management, supervisory control, and data acquisition services.

The multi-stage hydraulic fracturing design depends on the local geology and the nature of the reservoir. The design specifies the type of well completion, the number of fracture stages, fluid volumes and type, ratio of fluid additives, and injection rate to achieve the desired height, width, length, and complexity of fractures. As fracturing proceeds, operators adjust this design as more information becomes available. Depending on the reservoir, it may be preferable to fracture with a non-aqueous fluid such as gelled propane/butane, liquid carbon dioxide, or nitrogen to avoid exposing the shale to water. Only experienced companies can provide these more exotic and more costly fracturing fluid treatment services.

Shale Gas Technology and Well Integrity

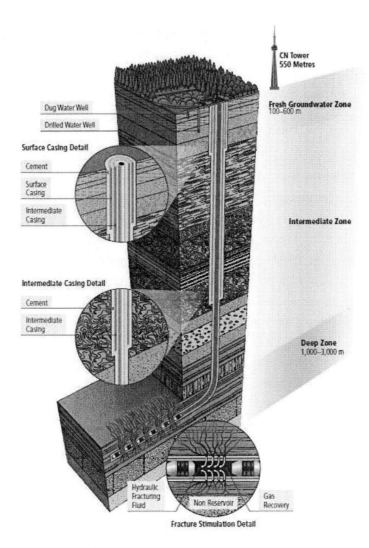

Figure 3.3. Well Construction Diagram for a Shale Gas Well. Schematic of a shale gas well, illustrating the various geological layers through which a well is drilled and the relative depth at which hydraulic fracturing occurs. Some laterals (the horizontal part of the well) are much longer than shown in this diagram and can reach up to 3 kilometres. The first two insets show the various casings (the steel tubing) that are inserted into the well and cemented into place. The bottom inset highlights a stage, a section of pipe between two packers that has been perforated in order to inject the hydraulic fluid to fracture the shale.

Depending on local geological conditions, the horizontal well section may be from 1.0 to 3.0 kilometers long. Because it is difficult to maintain a high enough injection rate to fracture the shale surrounding the entire horizontal well length efficiently through multiple ports in a single operation, the fracturing is done in stages, usually starting from the well toe (i.e., farthest from the wellhead) and moving toward the well heel, using one of several different methods. One approach involves perforating a limited length (5 to 15 metres) of the well casing at regular intervals, executing a fracture over a limited section of the well (called a stage), then installing a drillable packer to isolate the fractured section, perforating another interval closer to the heel, and repeating the process. The distance between fracture stages depends on local conditions or the operator's preferences, but typically ranges from 100 to 300 metres (ALL Consulting, 2012; and Rivard et al., 2012) with up to 15 to 30 stages along the length of the horizontal part of the well. The injection time for each fracturing stage varies between 20 minutes to more than 4 hours (King, 2012). The specifics of the fracturing treatment and the changes during the operations (e.g., from gelled water and proppant sand to water with friction reducers) will vary with the operator and the formation; they are also evolving rapidly.

The noise levels during the injection phase are substantial: between 4 and 24 fracturing trucks are simultaneously operating at top output and other engines operate blenders and pumps. During the high-pressure injection phase, workers are not permitted near the trucks, which are all manifolded and remote controlled. Rigorous safety standards are enforced.

Regulatory restrictions on executing multi-stage hydraulic fracturing near faults, near legacy wellbores (active or inactive), and at depths that are close to the base of the Fresh Groundwater Zone (FGWZ), vary from jurisdiction to jurisdiction. They are intended to reduce the probability of any fracturing fluid escaping the treatment zone during and after the high-pressure period and impairing other facilities or causing environmental degradation. However, as instances of unintended well-to-well communication illustrate, models and data bases are limited by ignorance of nearby active or abandoned wells that may provide a pathway for injected fracturing fluids past tight caprocks.

The propped zone, combined with the zone of shear dilation that surrounds it, is called the *stimulated reservoir volume* (SRV).[5] It is shaped like an ellipsoid that has grown upward more than downward from the injection point, with the details of its shape being a complicated function of the natural stress field, the natural fracture and bedding fabric of the rock mass, and the strategy used during the hydraulic fracture stimulation. An analysis of microseismic data of 12,000 hydraulic fracture stimulations indicates that induced fracture heights above any horizontal wellbore are limited by the volume of hydraulic fracturing fluids injected (Flewelling et al., 2013).

Monitoring to enhance the effectiveness of the gas extraction process involves tracking all injection parameters (e.g., rate, pressure, compositions, temperature, and density) continuously — particularly in the first wells. Microseismic data are used to delineate the extent (width and height) of the stimulated zone so that future multi-stage hydraulic fracturing operations in the same field can be designed efficiently and avoid wasteful out-of-zone fracturing. Deformation monitoring, less widely used, can inform about the shape and extent of the zone by measuring minute changes in the inclination of the ground during fracturing and flowback (Lolon et al., 2009). After the well is producing, performance data are collected and optimization analysis is performed to help improve treatment strategies for other wells in the region.

Hydraulic fracturing operations consume significantly more water than do conventional natural gas operations (King, 2012; and Rivard et al., 2012) though the amount needed depends largely on the geology of the play. The thickness of the shale, in particular, governs the type of fracturing and the pressure needed. Each fracturing stage can consume several hundred to a few thousand cubic metres of water, with an average of about 15,000 to 20,000 cubic metres for each well (Rivard et al., 2012).

---

[5] Installation of behind-the-casing pressure sensors during primary cementation would be a highly valuable source of information about gas seepage, but this is rarely done, and more rarely published.

The fracturing fluids used most commonly are *slickwater*, water with a viscosity-reducing agent such as polyacrylamide to allow the fluids to travel further into the rock fractures with lower pressure losses, and gelled (viscosified) water, which helps carry proppant into the rock mass (King, 2012). In many cases, both are used sequentially in each fracture stage to maximize both the proppant penetration and the stimulated volume. However, geological conditions sometimes preclude the use of slickwater. *Energized fluids*, the alternatives to water-based fracturing fluids, include propane-or butane-based liquefied petroleum, carbon dioxide and nitrogen gases, or foams. Gas-based hydraulic fracturing reduces recovery time and creates less formation damage but is more expensive. Propane is also flammable, making the treatment slightly more dangerous.

While the proportion of chemical additives in slickwater fracturing fluids is typically small — about 1 or 2 percent by volume or less — the quantities of water required for most fracturing operations can lead to significant amounts of the chemicals being used. One percent of a 50,000 cubic metre hydraulic fracture stimulation, for example, would be 500 cubic metres of chemical additives. If there are 10 wells on the same well pad, all undergoing the same treatment, 5,000 cubic metres of chemical additives would need to be transported to this single site, representing up to 200 truck trips for the chemicals alone over a 50- to 80-day fracturing period (although recycling[6] could reduce this volume significantly). Most of the chemicals are non-hazardous — guar gum, a naturally occurring polymer, or potassium chloride, which reduces formation damage — but given that only a few micrograms per litre of some additives could contaminate drinking water, the sound management of these chemicals at the surface is essential to protect both human health and the environment. Less toxic alternatives exist in some applications, but many are currently more expensive or less effective. The industry has, however stated that they will continue to develop safer alternatives for use in fracturing fluids.

---

[6] In shale gas development, recycling refers to the use of flowback water in subsequent fracture stages or transport to another well pad for reuse, thus substantially reducing the amount of freshwater used for hydraulic fracturing.

Once a hydraulic fracturing stage is completed, the injected fluids are allowed to flow back until the next fracturing stage begins. Flowback of fracturing fluids occurs slowly and at a diminishing rate as the well is producing. Before actual commercial production starts, the natural gas that accompanies the flowback fluid can be collected, vented, or flared from the first few wells while production rates or the right size of connecting pipelines are determined. Flaring is preferred if the natural gas flowback during the well completion operation cannot be recovered.

Once a single fracture stage is finished, the necessary modifications are completed so as to move onto the next stage, and the fracturing process is repeated until the well is completed. The next well on the pad is then rigged up and fractured.

Flowback fluid is a combination of the returning hydraulic fracturing fluid and water from the formation (salinity between 10,000 mg/L and 100,000 mg/L total dissolved solids). In addition, the flowback may also contain naturally occurring radioactive materials (NORM), metals, and organic compounds derived from the shale formation (Rivard et al., 2012). As a general rule, 30 to 40 percent of injected fluids will be recovered from hydraulic fracturing, although it can be much greater or much less (King, 2012; and Rivard et al., 2012).

Some areas are geologically suited to massive deep wastewater disposal because thick permeable saline aquifers are available at reasonable depth (0.5 to 1.5 kilometres). Regions generally do not have strata that would permit deep aqueous fluid disposal. In these cases, treatment and recycling are preferable.

The industry goal is to reuse all flowback water during subsequent hydraulic fracturing stages. Perhaps 5 to 10 percent of the flowback water will end up being deep-well injected because of its salinity or treated for release into surface water. The remainder will be recycled. Because of its high salinity and the presence of divalent cations, flowback water often needs treatment or more additives before it can be reused and is generally less effective for fracturing than fresh water (Rivard et al., 2012).

Technical and economic limitations influence the degree of feasible recycling in different areas (in some cases, for example, not enough fluid flows back, or flows back too slowly, for recycling to be viable). The trend towards more recycling is fostered by the high cost of treating or storing flowback water and the evolution of the treatment technology. Where many wells are drilled in close proximity, temporary pipelines and centralized treatment plants for the flowback water are feasible.

Among the advances for minimizing the use of fresh water for hydraulic fracturing is the use of saline groundwater. Saline water in sufficient quantities is readily available from productive aquifers in some but not all regions. For saline water to be effective, it may be necessary to increase the amount of additives such as viscosifying agents, leading to higher costs.

Neither propane nor methane fracturing requires water. Propane can be gelled into liquid form to transport proppants but methane cannot. Neither of these approaches nor liquid carbon dioxide and nitrogen fracturing, cause formation damage (e.g., swelling of formation shales), which aqueous fluids do.

Generally, flowback water is stored in tanks rather than ponds at the well pad. In general, ponds are only used to store water for fracturing liquid or non-oil-based drilling muds; oil-based muds must be stored in tanks.

3.2.2. Fracturing Chemicals

Chemical substances and proppants are added to the fracturing fluid at different stages of the fracturing process. The majority of additives in slickwater fracturing are used to reduce the viscosity of the fracturing fluid so that it will more easily penetrate the existing natural fractures in the formation. At first, the fracturing fluid carries only friction reducers, clay stabilizers, or other additives to facilitate its flow.

Once the fractures in the shale are created, granular material (usually sand) is added to prop open the fractures. Gelling agents (guar or xanthate gum) are added at this stage to increase the viscosity of the fluid to carry the proppants into the fractures. Breakers in the gelling agents activate after the proppants have been embedded in the fractures, causing the gel to liquefy, promoting flowback and recovery of some of the fracturing fluid at the surface, and subsequently allowing gas or oil to flow through the induced and natural fractures. These proppants represent 6 to 9 percent of the total injected volume with the chemical additives representing 0.5 to 2 percent (ALL Consulting, 2012) (Figure 3.4).

Multi-stage hydraulic fracturing is a rapidly evolving technology and many companies prefer procedures that differ from the sequence described above. A great deal of experimentation is taking place to optimize multi-stage hydraulic fracturing in various geological environments, and no single procedure will evolve as being best for all reservoirs.

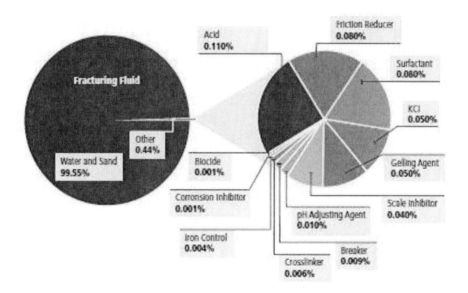

Figure 3.4. Fracturing Fluid Composition. The ratio of the different components that typically make up slickwater fracturing fluid. Water and proppant (sand) make up the majority of the fluid, with the remaining composed of a variety of chemicals.

The more common substances and chemicals used in hydraulic fracturing are listed in Table 3.1. Many of these have been used in the oil industry for several decades.

The additives used in fracturing fluids are generically similar (see Table 3.1), but each service company uses a different mix of chemicals based on site conditions, local geology, and company experience. These recipes, developed over several years, may represent a competitive advantage and companies are reluctant to reveal all aspects of their formulae (disclosure rules related to fracturing fluids are discussed in Chapter 4).

As mentioned, whereas the percentage by weight of chemicals used in the fracturing fluid is small, the absolute quantities can be substantial given the volumes of water used. For example, according to King (2012) in 20,000 cubic metres of viscosified fracturing fluid used for propping open fractures, there is approximately one and a half million kilograms of proppant, 100 cubic metres of acid, 1,000 kilograms of friction reducer, 900 kilograms of disinfectant and 0.3 cubic metres of corrosion inhibitor.

Table 3.1. Additives Used in Slickwater and Gelled Treatments

Additive Type	Purpose and Description	Common Additives
Water	Fresh water (less than 500 parts per million total dissolved solids)	-
Proppant	Maintains fracture openings to allow the flow of gas	Sand Clay or alumina ceramics
Friction Reducer	Reduces friction pressure, which decreases the necessary pump energy and subsequent air emissions	Non-acid form of polyacrylamide Petroleum distillate Mineral oil
Disinfectant (Biocide)	Inhibits the growth of bacteria that can destroy gelled fracture fluids or produce methane-contaminating gases	Glutaraldehyde 2,2-dibromo-3-nitrilopropionamide
Surfactant	Modifies surface and interfacial tension and breaks or prevents emulsions, aiding fluid recovery	Napthalene Methanol Isopropanol Ethoxylated alcohol
Crosslinker	Used for gels that can be either linear or cross-linked. The cross-linked gels have the advantage of higher viscosities that do not break down quickly	Borate salts Potassium hydroxide

Additive Type	Purpose and Description	Common Additives
Scale Inhibitor	Prevents mineral deposits that can plug the formation	Polymer phosphate esters Phosphonates Ethylene glycol Ammonium chloride
Corrosion Inhibitor	Prevents pipes and connectors rusting	N,N-dimethylformamide Methanol Ammonium bisulphate
Breaker	Introduced at the end of a fracturing treatment to reduce viscosity and release proppants into the fractures and increase recovery of the fracturing fluid	Peroxydisulphates Sodium Chloride
Clay Stabilization (e.g., KCl)	Prevents the swelling of expendable clay minerals, which can block fractures	Potassium chloride Salts (e.g., tetramethyl ammonium chloride)
Iron Control	Prevents the precipitation of iron oxides	Citric acid
Gelling Agent	Increases the viscosity of the fracturing fluid to carry more proppant into fractures	Guar gum Cellulose polymers (hydroxyethyl cellulose) Petroleum distillates
pH Adjusting Agent	Adjusts/controls the pH to enhance the effectiveness of other additives	Sodium or potassium carbonate Acetic acid

Data Source: Arthur et al., 2008; NYSDEC, 2011; and King, 2012.

3.2.3. Proppant Sand

Commercially produced sintered aluminum oxide beads, hollow glass beads, special fibres, and other materials may be used in multi-stage hydraulic fracturing proppant. For example, in deep applications, the fracture closure stresses are large during depletion, so companies may prefer the stronger aluminum oxide beads to quartz sand to prevent fractures closing.

Nevertheless, the great majority of the propping agents used are high purity well-rounded quartz sand (*frac sand*) carefully sieved and provided in bulk in a narrow range of grain size. Large sand grains are difficult to transport long distances into the fracture network during injection, but small sand grains are less effective in maintaining a large conductive fracture aperture. Different companies may specify different approaches to the amount, concentration, and staging of the frac sand grain size, but most are between 400 and 800 micrometres in diameter (0.4 to 0.8 millimetres or from 20 to 40 mesh size of standard screens).

Figure 3.5. Sand Mining Operations. An aerial view of a sand mine in Wisconsin. This mine produces frac sand that is used as proppant in slickwater fracture fluid.

Environmental (land use, water needs) and health issues (silicosis, industrial accidents) associated with frac sand mining, processing, and transportation are well understood as they are common to most surface mining operations and silica sand beneficiation.

3.3. WELL INTEGRITY

Striving for a high degree of well integrity to prevent immediate and longer-term leaks of gas and other fluids to groundwater or the surface is a cornerstone of environmental protection in any oil and gas drilling operation. The U.K. Royal Society and Royal Academy of Engineering (2012) states: *"Ensuring well integrity must remain the highest priority to prevent contamination."*

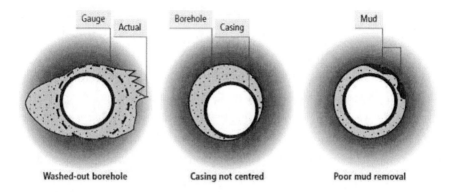

Figure 3.6. Cementation Issues. Poor primary cementation leading to potential loss of wellbore integrity may occur for several reasons: in the first, poor hole conditions and incomplete cement displacement; in the second, eccentric placement and incomplete cement entry; in the third, inadequate hole cleaning and mud-cake removal. In all cases, these situations may prevent a full cement seal from being achieved around the casing.

Several challenges must be overcome during the cementing process to prevent fluid (and gas) migrating into the surrounding environment either soon after cementing the well or years or decades later. Enough cement must be used to make sure it reaches an appropriate depth, covers the entire well casing, and displaces all the mud in the space between the casing and the borehole. In addition, the cement must be distributed over the entire length of the casing (i.e., no gaps, adequate thickness to prevent it from cracking), and it must be properly bonded to the steel casing and the rock. Gas can migrate into the cement while it is setting, which can also affect the integrity of the well. The casing must also be centralized in the borehole during cementing to make sure the drilling fluid is completely removed and the cement penetrates around the well casing (API, 2009).

Well drilling usually causes microfractures in the rock surrounding the well, and larger-scale damage such as washouts or massive breakouts may occur, especially in more intense stress fields (Hawkes et al., 2004).

Depending on the nature of the geological strata, this localized damage along the wellbore may offer or enable a pathway for slow upward gas migration as the cement for the annulus seal is generally not expected to fully invade the microfractures in the damaged zone because of the large grain size of the cement powder (Eklund, 2005).

Poor well construction affects the oil and gas industry as a whole and is not unique to shale gas. However, the problems are amplified by the potentially high number of wells associated with large-scale shale gas development and the chemical additives in fracturing fluids. This is a long-standing engineering challenge (see Jackson et al., 2013b for a discussion of the history of well integrity) identified by the oil and gas industry as early as the 1970s (e.g., Cooke Jr., 1979) and demonstrated in relatively recent shale gas wells as surface casing gas venting from the production zone and the Intermediate Zone (see Muehlenbachs, 2012a).

Achieving high-quality casing cementation is universally acknowledged as more challenging for inclined casing (e.g., a horizontal well) than for vertical casing, and particularly in that part of the well that has a sharp radius of curvature as it transitions from vertical to horizontal over several hundred metres. During the cement placement, it is difficult to keep the casing properly centralized in the borehole, despite using numerous downhole centralizing devices, and consequently uncemented space can result.

Other circumstances that can lead to inadequate casing cementation include autogenous cement shrinkage (a natural process in the curing of medium density cement slurries), improper cement formulation, and incomplete drilling fluid displacement (Dusseault et al., 2000). Cement may crack, shrink, or become deformed over time, thereby reducing the tightness of the seal around the well and allowing fluids and gases — often gas from intermediate, non-commercial formations — to escape into the annulus between casing and rock and thus to the surface (Dusseault et al., 2000). In addition, and all other things being equal, the challenge of ensuring a tight cement seal will be greater for shale gas wells that are subjected to repeated pulses of high pressure during the hydraulic fracturing process than for conventional gas wells. This pressure stresses the casing and therefore the cement that isolates the well from surrounding formations repeatedly (BAPE, 2011b).

Historically, have not required well integrity tests once casing is placed and the well is perforated and fractured; such tests are only done after the well is drilled and before the fracturing process starts (Precht & Dempster, 2012). Therefore, a cement sheath that may have been damaged by the completion or production phase is generally not identified, increasing the risk of future leakage behind the casing.

Operators can use a number of tools to check the cement integrity of their wells:

- Pressure tests that are restricted to localized sections of the casing and therefore test only the integrity of the steel casing across the isolated interval and not the condition of the wellbore behind the casing.
- Devices that, acoustically or otherwise, detect flow between formations behind the casing provide the most definitive results but are also expensive and cannot detect slow seepage of gas.
- Cement bond logs, which are relatively unsophisticated and inexpensive, have been standard practice for decades. They can however yield ambiguous results, and the resulting data cannot be easily linked to a direct estimate of leakage risk. A new generation of cement evaluation logs is widely available.

More advanced technologies exist but are more expensive and have not yet been subject to independent performance assessments.

The issue of well integrity applies not only to new shale gas wells but also to existing and abandoned conventional wells as any of these can offer or even develop a conduit for gas seepage. This issue is particularly relevant in areas with an extensive oil and gas drilling history and where standards ensuring well integrity were historically lower than they are today.

The impacts of these leaking wells are unknown because adequate characterization and monitoring have not been done. Mathematical models predicting the leakage and long-term cumulative impacts are unreliable due to the uncertain parameter inputs required, simplifying assumptions, and lack of field data for verification.

In addition, monitoring at the wellhead using conventional methods does not definitively determine gas leakage because gas may be leaking into underground aquifers. Surface casing gas venting at the wellhead does not indicate if any gas has migrated behind the casing to the near-surface and is being vented even a few meters away from the wellhead.

The oil and gas industry has substantially improved on the practices used for cement sealing wells over the past decade, and there is no doubt that the cements used today are much more effective. However, the degree of improvements claimed has not been independently tested or verified. A continuing effort is required to improve methods of wellbore cementation and cement logging with better tools, materials, and standards of practice. This needs to be coupled with greater due diligence by operators in identifying nearby faults and both active and abandoned off-set wells before starting drilling operations. Given that cement seals can be expected to deteriorate over time, repeated testing over long timeframes and appropriate mitigation requirements would be needed to maintain wellbore integrity.

3.4. GAPS IN KNOWLEDGE AND SCIENTIFIC UNDERSTANDING

Even when industry best practices are used, deficiencies in the methods determining the degree of well integrity remain. The results of geophysical logging methods (bond logs) and gas leakage measurements (e.g., surface casing vent flows, noise logs to detect behind-the-casing flow) lead to substantial uncertainties about the nature of the leakage pathways and gas leakage rates. There is currently no implemented method that satisfactorily provides such data.[13]

For the great majority of wells undergoing hydraulic fracturing, the maximum height of the upward propagation of the induced fracture is not well known because microseismic monitoring, or other methods to determine this, are often not applied. However, the data that are available suggest that this may not be a serious issue (Fisher & Warpinski, 2011). Nevertheless, high-quality data collection and interpretation needs to be placed in the public domain in the case of the very large fracturing operations taking place.

CONCLUSION

This chapter summarizes information about the key steps in shale gas development, focusing on the various technologies involved. These technologies have improved markedly over the past decade but are not risk-free: a continuing effort is required to improve wellbore cementation for example.

Several important uncertainties concerning the environmental implications of these technologies became apparent. In brief, these uncertainties relate to the:

- absence of important baseline information about both geological and environmental conditions in shale gas regions;
- performance of key components of shale gas development technology;
- pathways, fate, and behaviour of industry-related contaminants in groundwater;
- rate and volume of fugitive methane emissions;
- cumulative effects of development on communities and land; and
- risks of human exposure to industry-released chemical substances.

While these uncertainties arise in most cases from a paucity of relevant information, uncertainty flows in others from the very nature of the issue: the undefined scale and pace of future development, substantial regional differences, continued technical advancements, and the difficulty inherent in anticipating impacts far into the future.

Chapter 4

WATER

This chapter examines the current state of knowledge on shale gas development and water resources. The emphasis is on groundwater because the impacts of shale gas development on this aspect of the hydrological cycle are the most uncertain and controversial and the subject of the vast majority of peer-reviewed studies. The literature on groundwater impacts has grown markedly in the past three years, though the data are generally limited and commonly do not support definitive conclusions.

This chapter begins by presenting the framework used to consider the literature, establishing terminology, and describing the main features of the groundwater system. Next, subsurface pathways for potential migration of contaminants are conceptualized. The impacts of natural gas on groundwater quality and related assimilation processes are also identified. Surface water concerns and water use issues are reviewed along with challenges posed by the storage, treatment, and disposal of hydraulic fracturing flowback water. Lastly, gaps in knowledge and scientific understanding are summarized. This chapter only notes deficiencies in groundwater monitoring. A scientific framework as well as objectives and methods for monitoring are outlined in Chapter 8.

Groundwater can be considered in the context of the three zones illustrated in Figure 4.2. The Fresh Groundwater Zone (FGWZ) has potable water and somewhat deeper water that can be made potable by minimal water treatment.[7] The natural water chemistry in the FGWZ varies greatly between regions and even within regions depending mainly on the nature of the groundwater flow system and the geological formations through which the water flows. Identifying shale gas impacts therefore poses different challenges from region to region.

The term *Deep Zone* has no formal scientific meaning. It refers to zones with shale beds that are targets for gas extraction by hydraulic fracturing. Between the Deep Zone and the FGWZ lies the *Intermediate Zone*, where the groundwater is brackish to saline and may contain formations with entrapped gases that have no commercial value and thus were bypassed during drilling and well completion.

Two sources of contaminants — sources at land surface and those from below the fresh groundwater resource — can pose a threat to the quality of shallow fresh groundwater and surface waters. The main pathways by which anthropogenic chemicals can enter natural waters are from accidental surface releases at shale gas pads where the chemicals are stored and used in operations, or along transportation routes. Another way that natural contaminants and hydraulic fracturing chemicals could contaminate the FGWZ is from the flowback water that arrives at the surface during and after hydraulic fracturing. This water is handled and stored (and often recycled) at the pad and only removed when hydraulic fracturing operations are completed. This water is very saline and laden with hydraulic fracturing chemicals and other types of natural contaminants.

The oil and gas industry assesses impacts on groundwater as a result of the exploration, production, distribution, and use of petroleum products in two categories: upstream impacts and downstream impacts. *Upstream* impacts are those associated with the exploration stage and then the production stage in the field. *Downstream* impacts are those associated with refining and distribution, including leaks along pipelines and at retail stations. Examination of the publically available information indicates that the industry has researched and assessed upstream impacts on groundwater

---

[7] Shallow groundwaters in the FGWZ typically have low salinity of less than 1,000 mg of total dissolved solids per litre (1,000 ppm). However, the FGWZ may include brackish waters up to 4,000 ppm.

only minimally and that government and academia have also paid little attention to this category of impact. In some areas, the intensity of groundwater use may not peak until long after shale gas development has occurred. In most, but not all areas, shale gas development is too recent to produce clearly attributable contamination. For this reason, Jackson et al., (2013b) considered all upstream oil and gas activities as an indicator of what could likely occur with shale gas development because these activities have a lifespan of many decades or longer.

4.1. OVERVIEW OF POTENTIAL GROUNDWATER IMPACTS

A framework could be developed for considering shale gas impacts on groundwater according to a range of contaminants, approaches, methods, and groundwater conditions. The framework deemed appropriate because the range and toxicity of contaminant types that pose threats to groundwater from shale gas development exceed those for conventional oil and gas development. For example, a drilling site for conventional oil and gas development has fuels, lubricants, and a few other minor sources of contamination. Shale gas pads typically have these same chemicals as well as much larger amounts of the potentially more hazardous chemicals used for hydraulic fracturing. In addition, the flowback water contains a large number of natural contaminants derived from the shale rock. For the time that each pad has hydraulic fracturing operations, generally a few months or less, it is a storage site for hazardous chemicals. However, the waste (e.g., produced water) is not classified as such under existing regulations (e.g., B.C. Oil and Gas Commission, 2013c).

The goal of many regulations is to protect all groundwater resources suitable for societal uses, now and in the future. Consistent with the principle of sustainability, the goal goes beyond preventing contamination from most industrial, municipal, and agricultural activities of those water wells that exist in an area to protecting the aquifer itself.

Recalcitrant contaminants — those that do not effectively degrade by natural processes — in the shallow groundwater beneath a property usually end up contaminating groundwater beyond the property boundary, including public water resources, as all shallow groundwater is in motion. However, although recalcitrant hydraulic fracturing chemicals released at each well pad can be expected to migrate off-site, not all the effects are significant and some cannot be detected using existing methods of analysis.

None of the following types of commonly found industrial, agricultural, or municipal contaminants were anticipated decades ago:

- halogenated organic chemicals such as trichloroethylene and perchloroethylene from industrial cleaning;
- nitrate and pesticides from agriculture;
- food and/or pharmaceutical chemicals and pathogens from septic tanks;
- petroleum hydrocarbons including the gasoline components benzene and toluene; and
- road salt.

Because groundwater flow is slow, it can take decades or longer for contamination by recalcitrant chemicals to become a recognized problem. Furthermore, analytical techniques and monitoring of well networks were generally not sufficiently developed to detect emerging contaminants until a related public-health issue was clearly identified. In some cases, such delayed mitigation has resulted in widespread contamination. Therefore, considering the impacts of shale gas on groundwater must be framed in the context of decades or even centuries and anticipate potential effects that are not currently observed because evidence is not being sought.

Contaminants from shale gas development may impact groundwater via a number of potential pathways. When grouped by sources, contaminants include pad sources, non-pad infrastructure sources, and deep sources. Deep sources are anything originating below the FGWZ (i.e., either in the deep, gas-bearing target zones or in the Intermediate Zone, which may include both gas-bearing horizons and saline groundwater). While hydrogeologists know much about the FGWZ and petroleum geologists and engineers know much about production horizons in the Deep Zone, the Intermediate Zone has not been systematically studied to any degree other than for production of saline waters and hydrocarbons. Nonetheless, many of the technologies needed to study this zone do exist; they are used to understand the shallower and deeper zones and for occasional studies of the Intermediate Zone.

As described in Chapter 3, shale gas development requires building and operating an extensive temporary infrastructure as well as considerable truck traffic to bring water, chemical additives, and sand for hydraulic fracturing stimulation and oils and fuel to and from drill sites. In the case of slickwater stimulations (i.e., one in which the stimulating fluid is about 99 percent water), this infrastructure includes above-ground pipelines or roads to bring in the water and remove waste fluids, storage ponds or tanks, individual or centralized water treatment facilities and, where geologically feasible and legally authorized, injection wells to dispose of the surplus flowback water. The risks that this infrastructure and these operations pose to surface water and groundwater stem from:

- accidental spills of chemicals, oils, drilling muds, and fracture fluids during transportation, storage, or use;
- spills of condensates (where these are present) or flowback water from the producing well; and
- inadequate storage, treatment, or disposal of flowback water, which includes both fracturing fluids and saline formation water, and leaks from surface storage ponds or other storage facilities.

These are illustrated in Figure 4.1.

Threats from these sources can be greatly minimized by effective regulations. For regulations to be effective there must be performance monitoring and sufficient inspectors to ensure regulations are followed (Chapters 8 and 9). Although performance monitoring and enforcing regulations greatly reduce the risks of releases to groundwater, the risks are not entirely removed. The mobility, toxicity, and behavioural characteristics of the hydraulic fracturing chemicals in the flowback water therefore need to be understood to design an adequate monitoring system and mitigation measures.

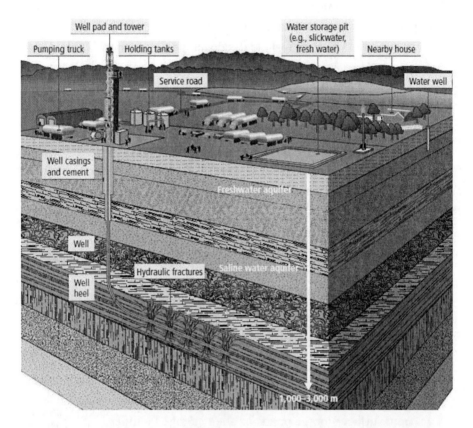

Figure 4.1. A Shale Gas Well Pad. Drilling a gas well involves the construction of a temporary drilling pad and pit to hold fluids and, in isolated areas, may also require the construction of an access road. The heavy equipment used in this construction includes: a derrick, transport trucks, pump trucks, lengths of pipe, and various tanks to hold chemicals and contaminated water. A single well may require some 2,000 one-way truck trips to bring all the supplies and equipment to the site. Note that the figure is a simplified illustration and does not include all of the infrastructure that would be at the surface of a shale gas well pad.

Of the wide range of views on the actual and potential impacts of shale gas development on groundwater quality, a common statement in the non-peer reviewed literature is that no impacts have been proven or verified Jackson et al., (2013b) provide a much more nuanced statement of this generalization: "There is no evidence that fracture propagation out-of-zone to shallow groundwater has occurred from deep (>1,000 metre) shale gas reservoirs, although no scientifically robust groundwater monitoring to detect gas migration has been attempted to our knowledge." That is, they do not rule out the potential for contamination in cases of shallow hydraulic fracturing.

Note also the distinction between contamination "directly attributable to hydraulic fracturing," as the AWWA stated, and the larger array of processes associated with shale gas extraction, which may also include wastewater reinjection and cross-contamination between Intermediate Zone layers and shallow groundwater due to poor or absent cement seals surrounding oil and gas industry wells. Vidic et al., (2013) summarize this controversy as follows:

Since the advent of hydraulic fracturing, treatments have been conducted, with perhaps only one documented case of direct groundwater pollution resulting from injection of hydraulic fracturing chemicals used for shale gas extraction. Impacts from casing leakage, well blowouts and spills of contaminated fluids are more prevalent but have generally been quickly mitigated. However, confidentiality requirements dictated by legal investigations, combined with the expedited rate of development and the limited funding for research, are substantial impediments to peer-reviewed research into environmental impacts.

A claim that shale gas developments have no impacts on groundwater needs to be based on generally accepted science including appropriate data obtained from the groundwater system using modern investigative methods. Moreover, because intense development in most shale gas plays has been taking place for less than 20 years, questions about the longer-term cumulative effects cannot yet be answered. Experience from other types of contamination shows that impacts on groundwater typically take decades to develop and become increasingly difficult to remediate. However, applying modern investigative methods at the initial stage of groundwater impacts should determine those that may become most significant later (see Chapter 8).

4.1.1. Contamination from Below the Fresh Groundwater Zone

The Intermediate Zone consists of strata of fractured shale beds interlayered with other types of rock such as sandstone, siltstone, limestone, and dolostone. The strata have a large range of porosities and horizontal and vertical permeabilities. This zone may also have gas-bearing formations, though these are usually too thin or too small to be of commercial interest. These strata are rarely subjected to detailed geological and geochemical characterization apart from occasional chemical and isotopic analysis of gases in drilling muds. Operators may even be unaware of their existence.

4.1.2. Gas and Saline Water in the Intermediate Zone

There is limited literature that examines the impacts of gas and saline water from the Intermediate Zone on the FGWZ. The emphasis is typically on potential impacts from the much deeper shale gas zones. However, once perforated by a well, gas and brackish or saline water from the Intermediate Zone could contaminate groundwater. The key question is whether, the Intermediate Zone can provide a pathway for fluids from the Deep Zone to communicate with the FGWZ, either directly through the wellbore or through existing or generated fractures or faults.

In addition, Vidic et al., (2013) stated:

[…] it has long been known that groundwater is salinized where deeper ancient salt formations are present within sedimentary basins, including basins with shale gas *where these brines are present at relatively shallow depths*. An important research thrust should focus on understanding these natural brine transport pathways to determine whether they could represent potential risk for contamination of aquifers because of hydraulic fracturing.

Thus, it is necessary to increase understanding of natural brine migration so as to evaluate brine mobilization and redistribution in areas of shale development.

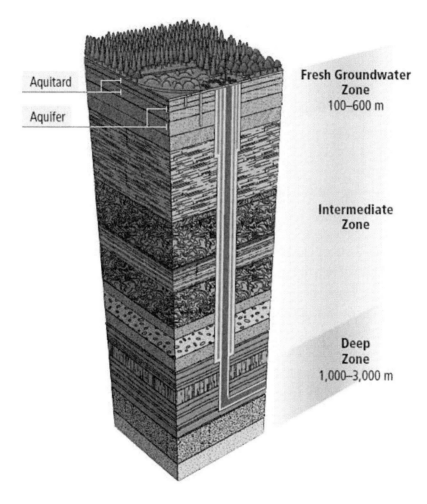

Figure 4.2. Depth Terminology. Schematic of a shale gas well illustrating the meaning of the terms Fresh Groundwater Zone, Intermediate Zone and Deep Zone.

Whereas the contaminative potential of the Intermediate Zone is likely much greater than that of the shale gas zone, the extent of fractures connecting to natural pathways or boreholes or seals has not been rigorously confirmed with field performance assessments and it likely varies by region.

The Intermediate Zone commonly has free gas with brackish or saline water in shale beds and also in other more permeable beds such as sandstone. The gas and the water are typically under full hydrostatic pressure; however, there may also be beds with excess or depleted fluid pressure (Zoback, 2010). If drilling results in pathways into or through the Intermediate Zone into the FGWZ, gas may migrate upwards along these pathways due to buoyancy or excessive pressure, potentially causing environmental impacts. However, the connectivity of these natural pathways is likely to be weak compared with pathways created by poor or missing cement seals along the wellbore.

As described in Chapter 3, there is always a risk that a cement seal in any particular oil or gas well, including shale gas wells, may leak in the future. In many jurisdictions the regulations do not require continuous cement seals through the Intermediate Zone (Chapter 9). Therefore, one of the most probable pathways for leakage is from the Intermediate Zone along the annulus between the cement seal and the rock into the FGWZ. Possible pathways are shown in Figure 4.3.

Existing cases of groundwater contamination due to upstream oil and gas activities have typically been caused by gas on account of its buoyancy and *in situ* pressure gradient. Brine or saline water are dense and not prone to migrating upwards along a well column or through fractured rock except from rare, over-pressurized zones. In some gas-producing regions, biogenic methane groundwater has been observed (Cheung et al., 2010; Osborn & McIntosh, 2010). Even thermogenic methane (e.g., Fountain & Jacobi, 2000) can be found close to the surface in areas without gas production. Biogenic methane generated in shallow aquifers and thermogenic natural gas from shales can be distinguished by a suite of geochemical and isotopic signatures. These include the relative concentrations of methane, ethane, propane, butane, and pentane, the stable carbon and hydrogen isotope ratios contained within these normal alkanes, as well as carbon-14, which allows discrimination between recent biogenic and fossil thermogenic hydrocarbons. In many cases, however, stray gas comes from several places, so isotopic tracing must be combined with other techniques and data to identify the sources of the leaking gas.

Water

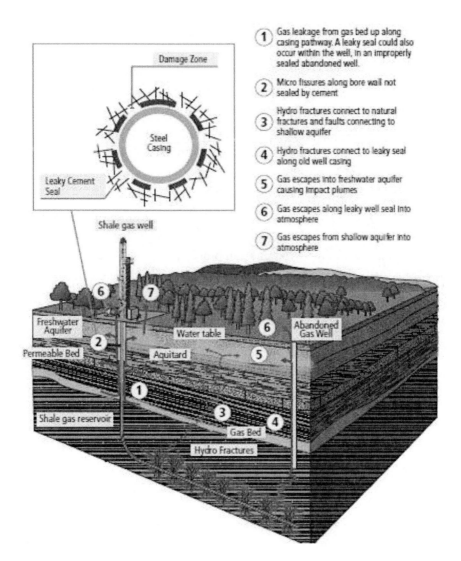

Figure 4.3. Conceptual Groundwater Contamination Pathways. There are several pathways by which potable groundwater could become contaminated by shale gas development, as shown in the schematic above. Note that this schematic is not to scale and does not imply that any of these pathways are necessarily present at any given site. The pathway marked by a dashed line is hypothetical as there is no known case of migration of hydraulic fracturing fluids from the deep shale zone to the groundwater level directly through the overburden rock.

These issues are not limited to shale gas and affect the whole gas-producing industry. They may, however, be exacerbated if the expected large numbers of wells are drilled for shale gas development.

Determining whether any methane found in shallow groundwater is the result of gas development activities or was already there can be difficult without establishing a baseline of dissolved and/or free natural gas using chemical and isotopic fingerprinting. Even when the depth of the source gas is clear, the pathway from the source to shallow groundwater is difficult if not impossible to discern because of the complexity of natural fracture systems and a lack of system characterization and monitoring to assess these systems. Furthermore, contamination may go undetected because of an absence of ongoing monitoring and sampling of domestic wells, dedicated monitoring wells, or other borehole sampling devices (Chapter 8).

4.1.3. Strata in the Intermediate Zone

A common misconception in some of the literature is that the Intermediate Zone typically has strata that are impermeable, such that they completely protect or isolate the FGWZ from the deep strata containing gas and saline waters. Any site-specific scientific evidence supporting this view from any shale gas region; the concept is contrary to ubiquitous geological heterogeneityc could not identify any. Effectively impermeable strata in the Intermediate Zone should be taken as the exception rather than the rule. Even if effectively impermeable rocks did exist in a shale gas environment, cross connection due to leaky well seals, abandoned wells, and fluid flow along faults could enable net fluid flow.

In horizontally layered geological formations with vertical hydraulic gradients, the layer with the lowest vertical permeability will control migration rates. However, investigating the Intermediate Zone to determine vertical permeabilities is technologically challenging. It is much easier to measure the horizontal permeability than the vertical permeability using existing test methods. In investigations conducted by the nuclear industry for deep geological nuclear-waste repositories, the vertical profiles of concentrations and isotope ratios of various dissolved constituents in the groundwater typically provide important information about the movement of fluids and solutes in the Intermediate Zone (Gautschi, 2001; Gimmi et al., 2007; and Clark et al., 2013). Therefore, methods for studying the mobility of fluids from shale gas formations upwards through the Intermediate Zone to the FGWZ do exist.

4.1.4. Natural Fractures and Faults as Pathways for Fluid and Gas Movement

One of the important characteristics of the groundwater environment that is particularly relevant to shale gas impacts on groundwater is the nearly ubiquitous occurrence of fractures in sedimentary bedrock. Fractures are generally the only pathways for substantial movement of water and gas through low-permeability rock and hence, in principle, into the FGWZ (Fountain & Jacobi, 2000). This migration occurs under natural conditions over geological time laterally along the bedding planes and upward through the joints. Although fractures are expected to close with depth due to overburden stress, this does not necessarily happen and naturally fractured, hydraulically active reservoirs are well known in petroleum geology (Bjørlykke, 1989).

Natural conduits are known to exist in shale rock. For example, natural faults in the Utica Shale represent a challenge to future drilling and fracturing (ALL Consulting, 2012). However, the mere existence of a conduit is not enough to contaminate potable groundwater as there also needs to be sufficient and sustained pressure to push the contaminating fluid to a height where it could overcome the hydraulic head of the freshwater zone. Most of the energy required to lift such fluids will be consumed in the process of shale fracturing and will not be available to drive a sustained flow of water to the shallow subsurface. Gas rather than brine and flowback water is the more likely cause of contamination of the FGWZ from below. Because they are buoyant and an upward gradient in fluid pressure is present, gases will behave differently here than saline water or hydraulic fracturing fluids.

Fractures in rock are mostly along the bedding planes, but there are also many others that connect the bedding planes, known as joints. These joints are typically oriented nearly perpendicular to the bedding planes, forming fracture networks that allow fluid movement. In a deep shale gas reservoir, the fractures are mainly closed due to the high stresses and lack of tectonic distortion (folding or bending). The gas therefore remains stored in these formations over geologic time because the leakage rate is extremely slow (Brown, 2000).

Faults are nearly planar narrow zones in the rock across which movement (shear slip) has occurred. These cross-bed ruptures broke the continuity of the sedimentary beds in the geological past so that the bed on one side of a fault is displaced from the same bed on the other. Many faults are more permeable than the surrounding rock mass and are therefore preferred pathways for fluid flow. Sealing faults, on the other hand, are made up of ground-up clay, rock fragments, or naturally deposited cements (e.g., calcite) that resist fluid flow and act as flow barriers. However, it is rare that faults in the Intermediate Zone are known to the extent that they can be classified as permeable or nearly impermeable.[8]

Hydraulic fracturing and other shale gas extraction activities may create or enhance preferential pathways for gas and saline waters to move upward more actively through the Intermediate Zone into the FGWZ. ALL Consulting (2012) presented a calculation for a worst-case scenario based on Darcy's law for upward flow of water from the shale gas zone. A recent literature review indicates that migration of hydraulic fracturing fluids and brine from deep shales through thick sedimentary basins is unlikely due to the constraining effects of low permeability formations (Flewelling & Sharma, 2014). However, the movement of the more buoyant natural gas through fractured sedimentary rock following its release by hydraulic fracturing has not yet been rigorously analyzed or assessed.

The greatest risk of gas or fluid migrating out of the production zone along existing faults and fractures occurs either during hydraulic fracturing when new flow paths are being opened and the formation is at the highest

---

[8] Studies have recently attempted to quantify this issue for the rock formations of the St. Lawrence Lowlands above the Utica Shale (Séjourné et al., 2013).

pressure it will experience, or if a well is shut-in (indefinitely closed) immediately after fracturing, allowing pressures to increase. Once gas production begins, pressure drops and gas and fracturing fluids tend to migrate towards the wellbore rather than to the surface by some undefined pathway.

The study of fractures in sedimentary rock, including their formation, properties, and role in fluid migration, is spread across several scientific and engineering disciplines, including geology, geomechanics, geophysics, hydrogeology, petroleum engineering, and petrophysics. Many aspects of fracture formation and properties remain ill-defined, and much is still unknown or poorly understood. This is especially the case for fractures in the Intermediate Zone because determining the nature of fracture networks here has been of little economic importance.

The importance of fluid migration in fractures on groundwater contamination was significantly underestimated at the outset for certain types of contaminants, such as chlorinated solvents (e.g., Kueper et al., 1992). Initial expectations for lack of migration were based mostly on supposition rather than rigorously acquired and analyzed data. In fact, most investigations of dense chlorinated solvents in fractured bedrock do not find the maximum depth of solvent penetration even after considerable (and expensive) investigative drilling because the maximum depth exceeds expectations. In addition, there is generally no practical benefit from determining the maximum depth beyond knowing it is large.

Nevertheless, deep, hydraulically active fractures can exist in sedimentary rock that has an overall small bulk permeability and their existence can be masked during hydraulic testing.

The accessible literature has not reported on any studies that measured the vertical hydraulic conductivity of shale beds or other relatively low permeability beds using borehole tests or other means. Such measurements are, however, undertaken at prospective nuclear-waste sites (e.g., Raven et al., 1992; Gautschi, 2001). Studies by Neuzil at the U.S. Geological Survey have demonstrated that many low permeability rocks do contain hydraulically active fractures that are largely responsible for the regional-scale permeability of these rocks (Neuzil, 1986, 1994).

Figure 4.4. Naturally Fractured Shales: The Marcellus and the Utica. Photos showing the natural fractures that can occur in shale rock. The top photo shows natural fractures in the shale (dark beds) and limestone (light beds) near Donnaconna, Quebec (Utica Shale). A pen is shown for scale. The bottom photo shows a drill core sample from West Virginia (Marcellus Shale) in which a calcite-filled natural vertical fracture is visible. These fractures are not necessarily representative of existing fractures at depth.

Nevertheless, deep, hydraulically active fractures can exist in sedimentary rock that has an overall small bulk permeability and their existence can be masked during hydraulic testing.

The accessible literature has not reported on any studies that measured the vertical hydraulic conductivity of shale beds or other relatively low permeability beds using borehole tests or other means. Such measurements are, however, undertaken at prospective nuclear-waste sites (e.g., Raven et al., 1992; Gautschi, 2001). Studies by Neuzil at the U.S. Geological Survey have demonstrated that many low permeability rocks do contain hydraulically active fractures that are largely responsible for the regional-scale permeability of these rocks (Neuzil, 1986, 1994).

Figure 4.4 depicts the geology of some shale gas formations at surface showing bedding features and vertical joints that can form fracture networks allowing flow in some situations.

Activities associated with shale gas extraction could enhance migration of gas through fractures or faults in two main ways. First, pathways for gas leakage may develop due to cross connections through parts of the Intermediate Zone to leaky wells. Second, activation of upward gas leakage along faults may occur as a result of minor earthquakes stimulated by hydraulic fracturing, or by slight distortions of the rock mass that allow these features to slip or open.

Seismological research shows that major earthquakes can increase the bulk permeability of bedrock and that this increase can last for up to a few years before the permeability returns to its previous state (Manga et al., 2012). During this time, movement of gas and saline water along fractures and faults is likely. Another potential result of hydraulic fracturing triggered earthquakes could be pulses of deep gas from the fracture networks that invade the FGWZ. These pulses would be related to the fluid pressure response in the fracture networks due to the earthquakes (Fountain & Jacobi, 2000). This topic is of particular interest geophysicists and needs greater attention with respect to shale gas production.

4.2. CHEMICALS USED IN HYDRAULIC FRACTURING

The shale gas industry has been criticized for its lack of transparency about the exact chemicals in the hydraulic fracturing additives. Many jurisdictions now require full or substantial disclosure (see Box 4.1). Some companies provide full disclosure voluntarily; others claim that the recipes must be protected as trade secrets. To assess any potential impacts and to design monitoring strategies, the exact chemical composition of the hydraulic fracturing additives as well as toxicity assessments and persistence and mobility tests are needed in surface and subsurface systems.

Apart from assessing each chemical on its own, the behaviour of these chemical mixtures and potential degradation products in waters under the expected variable *in situ* conditions including salinity, temperature, pH, and redox state for example, needs to be understood. There is only minimal reference literature and no peer-reviewed literature that assess the potential for the various chemicals in hydraulic fracturing fluids to persist, migrate, and impact the various types of subsurface systems or to discharge to surface waters. Van Stempvoort and Roy (2011) have identified those manufactured chemicals and fluids used in natural gas production in as well as the formation fluids (gases, saline waters, brines, and condensates) that are extracted. They also summarized what little information exists about the fate of various natural and manufactured chemicals in groundwater relevant to shale gas development operations.

Box 4.1. Fracturing Fluids

Fracturing fluids contain several different chemical additives that, depending on the operator and reservoir, are mixed in different recipes. In response to public concern about the risks that these chemical additives pose to human health and the environment, an increasing number of jurisdictions require disclosure.

Spills of hydraulic fracturing chemicals as a result of trucking accidents represent another potential source of water contamination in shale gas development. Although the conventional oil and gas industry handles much smaller volumes of water and chemicals, it shares with the shale gas industry many of the environmental risks for water contamination via surface pathways. Its environmental performance, therefore, provides a weak proxy for what one might expect in terms of surface accidents from large-scale shale gas development.

4.3. SUBSURFACE CONTAMINATION PATHWAYS

4.3.1. Vertical Fractures Created by Hydraulic Fracturing

The large volume of liquids used in a single shale gas well during fracturing means that the volumetric strain on the reservoir is an order of magnitude greater than in almost any previous conventional oil and gas well fracture treatments. This raises the concern that any induced fractures could breach the overlying geological strata and interact directly with shallow aquifers via existing faults and fracture zones (Myers, 2012,[9] Gassiat et al., 2013).

It is now understood that the volume of the rock mass that is affected by a fracturing operation can be far larger than the volume of rock reached by the proppant itself. This effect arises because the volumetric strains in the region close to the fracturing point cause stresses in the rock mass, and the high injection pressure reduces the frictional strength along natural joints. These processes lead to wedging open of more distant fractures and shear displacement across natural fractures. Because a natural fracture is a rough surface, if it is displaced by as little as millimetres, it will no longer fit together snugly when the active fracturing pressure dissipates during the flowback period. This *shear dilation* leads to enhanced flow capacity (i.e.,

---

[9] *High-risk non-compliance* is defined as one in which "a contravention of regulation(s)/ requirement(s) is found that the licensee has failed to address and/or that has the potential to cause a significant impact on the public and/or environment, and/or resource conservation" (Alberta Ministry of Energy, 2013).

68   *Afsoon Moatari-Kazerouni*

transmissivity) of the naturally fractured reservoir, opening up minute flow paths far from the proppant zone but still within the shale reservoir (Jackson & Dusseault, 2013). Industry has maintained that the risk of hydraulic fracturing creating vertical conduits that would communicate with, and therefore contaminate, shallow groundwater is extremely small for deep wells (i.e., those greater than about 1.0 kilometre). According to Fisher and Warpinski (2011):

Under normal circumstances, where hydraulic fractures are conducted at depth, there is no method by which a fracture is going to propagate through the various rock layers and reach the surface. This fact is observed in all of the mapping data and is expected based on the application of basic rock-mechanics principles deduced from mineback, core, lab, and modelling studies.

One unresolved issue is whether the volume changes in the shale gas zone as a result of injecting large volumes of liquid during hydraulic fracturing operations might bend or distort the overlying strata so that natural fractures in the rock open. Such deformation (rather than pressure) could generate new pathways for upward gas migration. Because the overburden rocks in many of the shale gas areas are stiff, small amounts of bending could be enough to open natural fractures even just a little, allowing naturally-buoyant gas to migrate upward.

This issue of strain magnitudes in the rock above the shale gas formation is one that can be studied quantitatively and subjected to mathematical modelling. However, verifying the stability of the hydraulic conductivity properties of the overburden during and after hydraulic fracturing requires sophisticated *in situ* strain measurements and long-term monitoring, neither of which has been done.

4.3.2. Existing Anthropogenic Conduits

Another potential pathway for groundwater contamination could be flow through existing anthropogenic conduits between a shale gas zone and the FGWZ. These conduits could include improperly abandoned oil or gas wells and old operating wells with faulty seals. Hundreds of thousands of improperly sealed, abandoned wells could exist in as legacy from conventional oil and gas production. Typically, this type of leakage occurs as gas flows up the annulus of the wellbore between the cement surrounding the casing and the rock wall exposed by drilling the borehole. Consequently, the most likely pathway for gas to seep from the Deep Zone and/or the Intermediate Zone to the FGWZ is via this pathway.

In many cases, multiple contaminant pathways may exist. Overall, the limited scale of studies that have detected thermogenic gas and other contaminants in drinking water wells near shale gas operations and the particular conditions in the study regions inhibit drawing firm conclusions about contaminant pathways. Even if baseline data did exist, it would not be possible to clearly differentiate contamination through natural pathways from that caused by previous or current drilling activities, leaky well casings, or from active fracturing. Without good baseline data, the task is immensely more difficult. Thus, in most cases, definitive claims in either direction can neither be proven nor disproven without better information from, for example, suitable science-based groundwater monitoring systems and from improved understanding of mechanisms of gas seepage from rigorous site-specific studies. This is discussed further in Chapter 8.

4.3.3. Faulty Practices during Drilling or Stimulation

Fracture treatments are generally closely monitored, but unexpected events in oil and gas exploration and production can lead to groundwater being contaminated by natural gas and/or fracturing fluids. In addition to accidental release at surface due to the activities already discussed, these can include:

- blow-outs of natural gas;
- hydraulic communication between the well-being fractured and a nearby production well (standoff well);

- shutting in a gas well before the annular gas is vented to the atmosphere or extracted and processed, which leads to pressurization of the wellbore; and
- direct injection of hydraulic fracturing fluids into the FGWZ or Intermediate Zone rather than the intended shale gas reservoir.

The near-surface groundwater was contaminated with large amounts of methane, leading to significant changes in the groundwater chemistry. The contamination caused elevated levels of dissolved ions (e.g., iron, sulfide), reduced levels of dissolved oxygen, and an increased pH of groundwater (Kelly & Mattisoff, 1985).

Cases of this kind are rare because regulatory authorities insist on the use of BOPs in drilling even in areas where natural gas is normally at low *in situ* concentrations. However, pressurized gas pockets that exist in the Intermediate Zone can cause unforeseen problems.

Only one documented case exists of a shallow aquifer becoming contaminated with hydraulic fracturing fluids, most likely as result of human error. This event took place during a stimulation of a shale gas reservoir and was due to the accidental injection of hydraulic fracturing fluids directly into sandstone at a depth of 136 metres when the operators believed they were fracturing at about 1.5 kilometres.

Although it may be unlikely that a hydraulic fracture in the Deep Zone would communicate with a shallow groundwater aquifer, induced fractures have been known to communicate with those around an adjacent well. The hydraulic fracture stimulation in the horizontal well caused fluids to discharge at the surface around the pump jack of the producing well (AER, 2012d). This type of communication can lead to the unintended discharge of water, gas, mud, or sands into FGWZ and Intermediate Zone aquifers and onto the surface.

4.4. THE ASSIMILATION CAPACITY CONCEPT

The important issue concerning groundwater impacts of shale gas development is not just whether such impacts occur, but whether these

Water 71

impacts become significant enough to be unacceptable. Groundwater monitoring is so rarely conducted that evidence of impacts is confined to cases of shallow well contamination (e.g., Gorody, 2012).

The concept wherein the groundwater zone has capacity to assimilate contaminants to purify the water, known as the *assimilation capacity*, is at the heart of any debate about impacts of shale gas activities, be they immediate or delayed. Claims of no impacts, nor of future accumulated impacts, must be based at least in part on the premise that the assimilation capacity for shale gas contaminants will not be exceeded. The assimilation capacity includes the concept of rejuvenating water quality within some acceptable distance or volume of aquifer over time. In other words, in favourable circumstances the groundwater system can be very resilient. The FGWZ can strongly attenuate many types of contaminants through a combination of degradation reactions, sorption, and hydrodynamic dispersion (mixing driven by mechanical dispersion and diffusion). These mechanisms occur over distances and timescales that vary depending on the contaminant and the characteristics of the hydrogeological system, if the loading does not exceed the FGWZ's capacity to assimilate the chemicals.

There are many examples of industrial or other activities that contaminate groundwater but for which the assimilation capacity prevents impacts or reduces them to acceptable levels. For example, contaminants such as polychlorinated biphenyls (PCBs) are effectively immobile due to sorption and therefore do not travel far enough as solutes in groundwater to do much harm, making their presence in the groundwater system insignificant in nearly all cases.

An example of exceeding assimilation capacity due to over intensification is illustrated by household sewage septic systems that produce acceptable, localized impacts, except where the geology, and the density of the systems, cause loadings that exceed the assimilation capacity. The same may be true of shale gas development as it intensifies in some areas.

Another example is leakage of gasoline from service stations. Because petroleum products are buoyant in their oily phase, they float near the water table. This shallow accumulation allows dissolved oxygen to become available for their biodegradation. Thus, the groundwater zone has demonstrated an immense assimilation capacity for refined non-halogenated petroleum products. However, if a contaminant plume derived from hydrocarbon fuels occurs in an area with many domestic wells, even such plumes of limited extent can harm. In this context, the degree to which the various chemicals in hydraulic fracturing fluids attenuate in freshwater aquifers is unknown but is likely variable due to the very different properties of chemicals used and the resulting variable mobilities.

Over decades, strong subsurface assimilation capacity may be the only factor that will prevent leakage from shale gas wells from substantially degrading groundwater resources. Although assimilation capacity is a key component in assessing the long-term impacts of shale gas development on groundwater, the literature contains no impact analysis that evaluates or even formally considers the assimilation capacity in the context of natural gas releases to the FGWZ.

4.5. IMPACTS OF NATURAL GAS ON SHALLOW GROUNDWATER QUALITY

As discussed in Chapter 3, incomplete or deteriorating wellbore seals can leak gas that comes from the Intermediate Zone or from the deeper shale gas reservoir. This leakage occurs either up along the interface between the cement seal and the formation between the casing and the cement, or through gas channels and pockets formed in the cement as a result of gas migration (Watson, 2004). Gas leakage pathways may result because of difficulties in positioning the cement or because the cement deteriorates over time. In many cases, there is no requirement to cement off thin gassy formations in the Intermediate Zone. Behind-the-casing pathways of gas leakage are often difficult to detect using standard geophysical logging tools (e.g., cement bond logs). Although improved logging tools are becoming available, they are fairly expensive to use and may not be required by regulations.

Water 73

This problem is not specific to the shale gas industry; many types of oil and gas industry drilling and production activities can impact groundwater quality. However, it is particularly relevant to shale gas development because of the relatively large numbers of wells that are drilled in the midst of domestic wells in rural and near-urban areas and because of the buoyancy of natural gas.

To assess the likely impacts of shale gas-derived methane contamination on the FGWZ, it is essential to fully understand the processes controlling its solubility and the geochemical and biogeochemical reactions that it may induce. Methane has limited solubility in water: approximately 23 milligrams per litre at 25°C or 28 milligrams per litre at 15°C when at one atmosphere (Gevantman, 2013). This solubility increases by 32 milligrams per litre per each 10 metre increase in depth. Dissolved methane readily exsolves upon pressure reduction making it difficult to determine saturation (Roy & Ryan, 2013). Thus, even relatively small decreases in hydraulic head (i.e., during abstraction/pumping) can induce exsolution and allow gas phase methane to fill the pore space (or head space of wells) leading to explosion hazards.

The issue to be addressed here concerns the potential impacts of this type of natural gas leakage on groundwater quality. An important associated issue concerns other water quality aspects, that is, the biogeochemical processes that may attenuate the gas during transport by groundwater in freshwater aquifers away from the leaky wells.

According to Vidic et al., (2013), "methane can be oxidized by bacteria, resulting in oxygen depletion. Low oxygen concentrations can result in increased solubility of elements such as arsenic and iron. In addition, anaerobic bacteria that proliferate under such conditions may reduce sulfate to sulfide, creating water and air quality issues." However, all of these impacts have not yet been confirmed in field investigations in areas of shale gas development.

The methods needed to assess the effects of methane contamination have been developed in studies of other types of groundwater contamination but have not yet been applied to assess the impacts of methane leakage from leaky oil and gas wells. Therefore, the degree to which the assimilation

74 *Afsoon Moatari-Kazerouni*

capacity of the groundwater zone for methane leakage will prevent long-term deterioration of groundwater quality remains unknown. The Van Stempvoort et al., (2005) study remains the best analysis (as well as the only refereed publication) of occurrence of methane and methane assimilation from an oil or gas operation into a freshwater aquifer. Furthermore, the significance of the aquifer as a buried valley aquifer is important in the context of groundwater supplies casing terminating in the shallow bedrock in order to draw water from multiple horizons at typical depths of 100 to 500 feet beneath ground surface") concluded that gas concentrations were best correlated with topography and groundwater geochemistry, and not shale gas extraction. The Duke University researchers followed up their study with a more comprehensive suite of indicator parameters that supported their previous interpretation (Vengosh et al., 2013). The various explanations are not resolvable in the context of methane migration pathways without improved characterization and monitoring, which means monitoring well systems rather than domestic wells, and perhaps sampling of domestic wells over longer time frames.

4.6. SURFACE WATER CONCERNS

Compared with groundwater monitoring, relatively little attention has been paid to monitoring surface water quality.

Discussing the Marcellus Play, Vidic et al., (2013) point out that:

It is difficult to determine whether shale gas extraction affected water quality regionally, because baseline conditions are often unknown or have already been affected by other activities, such as coal mining. Although high concentrations of [sodium], [calcium], and [chlorine] will be the most likely ions detected if flowback or produced waters leaked into waterways, these salts can also originate from many other sources. In contrast, [strontium], [barium], and [bromine] are highly specific signatures of flowback and produced waters.

Water 75

Thus, the situation with respect to surface water quality is not markedly different from that for groundwater quality; in both cases, baseline monitoring is inadequate.

Vidic et al., (2013) further observe that information about the quality of surface water is impeded by legally binding settlements between operators and landowners:

> Confidentiality requirements dictated by legal investigations combined with the expedited rate of development and the limited funding for research are major impediments to peer-reviewed research into environmental impacts.

This applies equally to understanding operators' groundwater contamination.

Increases in suspended solids due to the development on and around well pads have resulted in an increased sediment runoff yield (Williams et al., 2008). Additionally, Entrekin et al., (2011) studied streamflow turbidity in areas of Arkansas that are undergoing development of the underlying Fayetteville Shale. They noted a strong correlation between shale-well density and stream turbidity in seven drainage basins during the high flows measured in April 2009.

Heilweil et al., (2013) describe the development of stream-gas reconnaissance sampling to estimate methane releases as a viable means of evaluating groundwater impacts from unconventional gas development. The method involves measuring in-stream methane concentrations and groundwater discharge to the stream as well as modelling of in-stream mass transfer of methane.

While the extraction of water from surface water sources for shale gas production and the disposal of wastewaters are the main concerns, other concerns that relate to surface waters include impacts on hydrology from changes to the land; loss of buffer strips; habitat discontinuities associated with temporary road and culvert placement, maintenance and integrity; the potential consequences of dams or other structures associated with water collection or impoundments; increased sedimentation; and increased forest or aggregate resource use associated with improved access via transportation routes associated with shale gas development.

The magnitude of potential development, especially in remote areas, means that protecting surface water ecosystems and ecosystem services should be a priority.

4.7. WATER USE

The total amount of water needed for shale gas development is generally small in the hydrological context (i.e., relative to annual, total surface water flows). However, the hydraulic fracturing procedure requires large volumes of water over short periods of time (several weeks to months), which could create stresses due to quantity and related quality impacts at particular times of the year in some parts of the country. Problems may arise at the driest time of the year when demand is highest for many water uses, at the coldest time when surface waters are mostly frozen and active flow is low, or during critical periods when water levels are important for access to critical habitats.

It should be noted that the shale gas industry is trying to avoid such problems by recycling water, storing water on-site, using saline water, and even replacing water used in hydraulic fracturing with gas rather than water. However, the robustness of these alternatives and the degree to which the economics of shale gas will allow them to be applied is poorly understood. As noted in Chapter 3, shale gas development typically uses much more water than conventional gas development because of the added demands related to hydraulic fracturing. The amounts used, however, vary extensively from play to play, and sometimes even from well to well, depending primarily on the composition of the shale as well as a number of economic and other geological factors.

The very large volumes of water needed to hydraulically fracture shale gas wells with current technology make water consumption a critical issue in shale gas development. With

Water 77

hundreds of wells to be drilled over large shale gas plays, water management warrants considerable regulatory attention and could limit where, when, and how fast shale gas development occurs.

The source of the water used in hydraulic fracturing is primarily fresh water (e.g., lakes, streams, groundwater, or even municipally treated water). It can also come from deep saline aquifers and be reused produced water. Although more water is being recycled and reused, large quantities of freshwater are still required (provided the shales are not sensitive to low-TDS waters) as brackish water is more likely to damage equipment and lead to formation damage (Stark et al., 2012). Some jurisdictions may limit which sources of water can be used for hydraulic fracturing and, to reduce impacts, limit the volumes used and timing.

Accounting for up to 80 percent of total transportation activity, the haulage of large water volumes to and from drilling sites represents a significant logistical challenge and cost to industry as well as to the environment in the form of transport-related GHG emissions and air pollutants (Stark et al., 2012) (see Chapter 5 and 7).

The withdrawal of water for hydraulic fracturing represents a consumptive use of water since some of it will remain in the shale gas formation and not flow back to the surface. Is this consumptive use large or small? There is no straightforward answer to this question. Water use is large compared to conventional gas production, but small relative to the production of oil, particularly secondary oil recovery and production from the oil sands (on an energy equivalent basis).

The amount of water used in fracturing a shale gas well can also be compared to other economic activities.

These examples show that the absolute volumes withdrawn are often less important than the times and rates at which water is taken. Hydraulic fracturing uses a lot of water over a short period of time (several days). If several fracturing operations.

In most cases, however, this challenge can be managed by withdrawing water during peak periods (e.g., the spring freshet) and storing it until required. The recycling of flowback water and using non-potable water can also reduce the industry's demand for fresh water. All such options are currently being considered and their economic viability determined; it remains to be seen how regulations will affect their use by industry.

4.8. FLOWBACK WATER

4.8.1. Storage and Treatment

The flowback water that returns to the surface in hydraulic fracturing operations contains not only the chemical additives that were mixed into the fracturing fluid but also formation water that may be high in dissolved solids. As these dissolved solids are primarily salts, a flowback water leak or spill can increase the salinity of the receiving environment. The dissolved solids in flowback water may also include NORM and other natural components such as trace metals (e.g., arsenic, barium) that can contaminate water and accumulate on equipment, possibly posing a health risk to workers. TDS can range from brackish (<10,000 milligrams per litre) to saline (>100,000 milligrams per litre), creating a risk to the potability of freshwater if contamination occurs. Hydrofracturing fluid can also adversely affect or kill vegetation (Adams, 2011). Whereas the rates and chemical composition of flowback water vary significantly from region to region and the bulk of it is produced over the first few days after a fracturing operation, most conventional and unconventional gas wells will typically continue to produce small quantities of formation water while in operation (EPA, 2011b).

Flowback water is usually stored in lined surface ponds or tanks before being either treated on-site or off-site in a specialized treatment plant, reused to fracture another well, or reinjected into a deep saline formation. Lined ponds, even when built with double liners, are rarely free from flaws and can be expected to leak over time. Similarly, the permeability of clay-lined ponds can be increased by the salinity of the stored flowback water (Folkes, 1982).

Water 79

In addition, surface ponds can overflow as a result of significant precipitation (e.g., during heavy rain storms).

The challenges to rendering flowback water suitable for release into rivers or lakes differ substantially from those common to the wastewater treatment industry. The degree of difficulty, and hence the cost of treatment, depends on many factors including the salinity, the specific chemical composition of the fluid (including radioactivity), and the tolerance of the ecological system into which the treated water is to be discharged.

Saline fluids cannot be treated in typical municipal wastewater treatment plants because of their deleterious effect on the microbes in the activated sludge process. Similarly, NORM components may either be sorbed by the sludge or simply flow through the treatment plant and be discharged into receiving waters. Other treatment options include reverse osmosis and thermal distillation and crystallization (Gregory et al., 2011), some of which are being applied in treatment plants in Texas and Pennsylvania established specifically for treating flowback fluids. Warner et al., (2013) examined the composition of shale gas development effluent from a wastewater treatment facility in western Pennsylvania and of stream sediments up- and downstream from the discharge point. They concluded:

226 [Radium] levels in stream sediments (544 to 8759 becquerel per kilogram) at the point of discharge were [approximately] 200 times greater than upstream and background sediments (22 to 44 becquerel per kilogram) and above radioactive waste disposal threshold regulations, posing potential environmental risks of radium bioaccumulation in localized areas of shale gas wastewater disposal.

Minimal research has been conducted on this aspect of shale gas development. In addition, the costs of treating flowback waters to achieve ecological and human health and safety standards are generally very high with uncertain regulatory outcomes. Hence, deep-well injection is the industry's commonly preferred option when the geology is suitable.

80 Afsoon Moatari-Kazerouni

4.8.2. Deep-Well Disposal

The optimum practice in the oil and gas industry for disposal of wastewater is to inject it underground (MIT, 2011). Injection wells are sometimes shallower than production wells but still much deeper than freshwater aquifers. The disposal of waste fluids through deep injection is regulated wherever it occurs. Deep injection typically involves greater fluid volumes per well than is the case for hydraulic fracturing operations, albeit pumped at lower pressures. Waste fluids are injected into permeable porous formations that are specifically targeted to accommodate large volumes of fluid; they are often depleted oil and gas reservoirs or saline aquifers.

Deep disposal of wastewater poses two main hazards: risk of groundwater contamination and risks related to induced or triggered seismicity. The latter is addressed in Chapter 6. Deep-well disposal is a long-standing practice for disposal of saline fluids and acid gases in the oil and gas industry. The risk to the FGWZ should not be significant when best practices are followed because the low injection pressures and rates should not result in significant upward displacement through abandoned wells or leaky well seals.

4.9. Limits in Knowledge and Scientific Understanding

The different views in the literature on contamination pathways in the subsurface related to shale gas development indicate the need for more comprehensive and conclusive field research with supporting laboratory and modelling activities.

Several potential impacts of shale gas development are difficult to assess because of significant limits in current scientific understanding:

- Impacts on groundwater quality are generally not predictable using established scientific or engineering analyses because such impacts would likely be gradual, over decades or longer. Shallow leaks, for example, along wellbores or from surface spills to the FGWZ, would be detectable much earlier given a suitable monitoring strategy.

- The baseline or background hydrogeological and hydrogeochemical conditions of groundwater flow systems in the areas here there is, or may be, shale gas development are poorly understood.
- The behaviour of chemical additives used in fracturing in groundwater and their reactions with *in situ* fluids and rock are not well understood. The same is the case with attenuation of produced fluids released into fresh groundwater and surface water.
- The assimilation capacities of the groundwater zone for shale gas extraction contaminants — gas and hydraulic fracturing chemicals — are generally unknown and probably vary depending on hydrogeological environments. They need to be examined over appropriate times and distances to identify potential risks.
- The linkages between groundwater and surface water resources across the country are not well understood and historical surface water records for all of the areas under development are seldom good.
- Calculating minimum streamflows in rivers draining shale gas plays will require a consensus among stakeholders on limits of abstraction during low-flow periods.

4.10. CONCLUSION

Although there are published claims that no proven or verified impacts of shale gas development on groundwater exist, more recent publications and reports dispute these. The burden of proof should not be on the public to show impacts, but on industry to verify that their claims of performance are accurate and reliable over the relevant scales in space and time. There is reason to believe that shale gas development poses a risk to water resources, but the extent of that risk, and whether substantial damage has already occurred, cannot be assessed because of a lack of scientific data and understanding.

The main potential cause of groundwater contamination is expected to be from upward gas migration along well casings or in combination with natural fractures causing entry of gas over extended time into freshwater aquifers or into the atmosphere. In aquifers, the gas may be assimilated by natural geochemical processes, but these same processes may release natural contaminants such as metals and hydrogen sulfide that could degrade water quality. Rigorous baseline monitoring has not been conducted in potential shale gas regions, and the assimilation capacity of groundwater systems in these areas has not been assessed for contaminants associated with shale gas development.

Even if impermeable caprocks did exist above a shale gas reservoir, seepage via leaky well seals and abandoned wells and fluid flow along faults could bypass otherwise low permeability rock strata or displace fluids in the Intermediate Zone. The risks of such events are both variable and poorly quantified. They need to be carefully considered, particularly near wetlands, in populated areas served by domestic wells, and in near-urban areas that may have abandoned wells. Whereas gases migrate upward due to buoyancy or pressure gradients, hydraulic fracturing chemicals and saline waters are less likely to migrate upward from the hydraulic fracturing zone into the FGWZ because such an event would require sustained upward pressure gradients over many years.

The most important questions concerning groundwater contamination from shale gas development are not whether groundwater impacts have or will occur, but where and when they will occur, if they will occur to an unacceptable extent, and how long they will last. Finally, how and to what degree can regulations and industrial practice prevent such impacts, and how can impacts be mitigated or remediated once they occur?

Even though hydraulic fracturing uses substantial volumes of water, the total volumes are small relative to the existing water resources. Nevertheless, water use may be an occasional problem when the short-term demand of hydraulic fracturing competes with other water uses seasonally or in drought periods. These problems can be avoided by good water management practices that will necessarily include improved characterization and monitoring of drainage basins in areas of shale gas development.

Chapter 5

GREENHOUSE GASES AND OTHER AIR EMISSIONS

Chapter 4 addressed the implications that large-scale shale gas development raises for water quality and quantity, and described what was known about possible contamination pathways and water availability for hydraulic fracturing. Although public attention over the environmental effects of shale gas development has largely focused on these water issues, there are other environmental risks that need to be considered related to air and land, as well as human health. Issues related to GHG and other air emissions are addressed in this chapter, whereas those related to land impacts and stimulated seismicity are addressed in Chapter 6, and human health issues are addressed in Chapter 7.

5.1. GREENHOUSE GAS EMISSIONS

Shale gas is a fossil fuel, and its production and use lead to emissions of carbon dioxide and methane, both GHGs contributing to climate change. The environmental impact of shale gas with respect to anthropogenic climate change is not clear-cut. One recent review states that "estimates of greenhouse gas (GHG) emissions from shale gas production and use are controversial" (O'Sullivan & Paltsev, 2012). The British Geological Survey has concluded that "the overall greenhouse footprint of [...] shale gas, including direct and indirect emissions of both [carbon dioxide] and methane, is not yet fully understood" (Energy and Climate Change Committee, 2011).

How shale gas development affects climate change depends on its net contribution to global GHG emissions. Substituting natural gas for coal in electricity generation, for example, lowers carbon dioxide emissions per unit of energy produced, in part because of the greater efficiencies typically achieved in gas turbine power plants compared to coal-fired boiler power plants.

Compared with coal, and if the electricity generation function, is combined with a co-generation approach that uses part of the heat, it is claimed that close to 80 percent efficiency is achieved.

Natural gas development is acknowledged to have a far smaller land use impact compared to coal mining and thermal coal power plant siting, fewer impacts on safety and occupational health, and smaller pollution impacts in its production compared to coal.

Substituting shale gas for coal and liquid fuels can reduce the overall environmental impact and carbon footprint of primary energy use. These benefits may be reduced by methane leakage, discussed further below.

To estimate net contributions, it is necessary to perform a full *well-to-burner*[10] comparison of shale gas with other fuels that includes all sources of GHG emissions associated with the production, processing, transport, and consumption of each fuel. Such analyses vary considerably in their results, depending on the values chosen for variables including the GHG potency of methane (see below). Moreover, upstream, transportation, and storage leaks can reduce the benefits of lower emissions at the consumption stage, and the magnitude of these fugitive methane emissions is the subject of on-going scientific discussion. Other factors involved include the relevant time frame being considered; the impact of decreased atmospheric sulphate aerosols if natural gas replaces coal; the assumed rates of methane leakage and venting

---

[10] Some authors have used the term life-cycle analysis; however, felt that well-to-burner analysis is the better way to reflect the linear value chain of fossil fuels.

in natural gas production; transmission and storage facilities; the efficacy of flaring of waste natural gas; and the gas end use. Venkatesh et al., (2011) have stressed that life-cycle analyses frequently do not acknowledge and address such uncertainties which, in some cases, may change not only the magnitude but also the sign of the expected effect. In their analysis, the probability of substantially reducing emissions by replacing coal with gas in electricity generation is almost 100 percent, whereas the substitution of natural gas as a transportation fuel carries a 10 to 35 percent probability of increasing emissions.

Concern has been expressed that abundant supplies of cheap methane could both boost overall energy demand and delay the development of non-fossil fuel primary energy technologies, while still allowing large amounts of GHG to be emitted (Schrag, 2012). Socio-political factors, including those related to trade or investment in other energy sources may also impact the net effect of shale gas development on climate change. The U.K. House of Commons Energy and Climate Change Committee (2011) recently concluded that lower gas prices driven by shale gas development "have the potential to shift the balance in the energy markets [...]" and quoted a submission from the U.K. Department of Energy & Climate Change stating that unconventional gas development could "[...] reduce the incentive for investment in the 'low-carbon alternatives required to meet longer-term emission goals." Their report also noted the risk of increased use of gas contributing to locking in to high carbon infrastructure (Energy and Climate Change Committee, 2011). The broader effects of shale gas development on energy demand and global GHG emissions depend, therefore, on a complex set of market factors and national and international policies that are outside the scope of this chapter.

The following section focuses on the GHG emissions associated with shale gas production and gathering (so-called *upstream* emissions). This is an area that features conflicting evidence and opinions, primarily around the issue of methane leakage at the production stage. Before reviewing this evidence, it is useful to note the relative impact of methane and carbon dioxide as GHGs.

5.1.1. Warming Impact of Different Greenhouse Gases

Compared to carbon dioxide, methane has a much stronger GHG effect, but carbon dioxide typically remains in the atmosphere almost ten times longer than methane. Different types of GHGs are usually compared using a conversion factor called global-warming potential (GWP). GWP compares the average impact of other gases and aerosols on radiative forcing (i.e., warming potential) with that of carbon dioxide over a defined period of time. If the impact is averaged over 20 years, however, the GWP value increases to 84. This means that over a time frame of 20 years, a given mass of methane will create 84 times the amount of radiative forcing compared to the same mass of carbon dioxide. Both the 100-year and 20-year GWP values for methane listed in the fifth report[11] of the IPCC represented an increase over those listed in the previous version (IPCC, 2007, 2013). While most estimates use the 100-year GWP value, some scientists have argued that considering the impact over 20 years (and therefore using the higher GWP average) is more appropriate given the need to reduce global GHG emissions significantly to maintain the global increase in average temperatures below 2°C and avoid tipping points in the climate system (Howarth et al., 2011; Wigley, 2011). However, Schrag (2012) argues that the 20-year time scale is too short as it puts the focus on the current emissions rate as opposed to total cumulative emissions, the latter being the primary driver of climate change. Studies have shown that cumulative emissions have a much greater impact than the rate of emissions (Allen et al., 2009; Matthews et al., 2009). Thus, reducing methane emissions over the next 20 years would only delay a temperature rise by a few years and have a much smaller impact than reducing cumulative GHG emissions (Schrag, 2012), assuming that no climate tipping points are breached.

Due to the complex interaction effects among GHGs and between GHGs and aerosols such as black carbon, other groups are arguing that the GWP

---

[11] IPCC, 2013 had been released and "accepted by Working Group I of the IPCC but not approved in detail." The GWP values quoted here do not include climate-carbon feedbacks.

concept should be replaced by a more precise forcing equivalent index (FEI) that would compare the effect of different GHGs in the broader context of the atmospheric conditions into which they are being released (Manning & Reisinger, 2011). However, there is no debate that in all cases, methane is a more powerful GHG than carbon dioxide, and unburned methane released to the atmosphere is a significant environmental risk because of its contribution to climate change. Figure 5.1 shows the effect that different assumptions about emission estimates and GWP of methane have when comparing the GHG implications of using natural gas versus coal for electricity generation. This figure demonstrates that a higher GWP or a larger percentage of methane emissions resulting from development reduces the benefit of methane relative to coal.

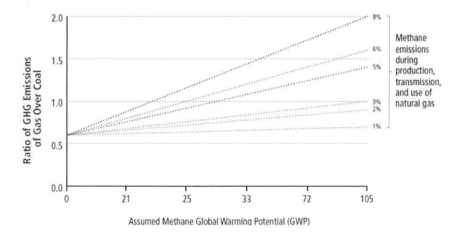

Figure 5.1. The Impact of Assumptions about Methane on the Comparative Well-to-Burner GHG Emissions of Natural Gas versus Coal. The impact that both methane leakage and the methane global warming potential (GWP) have on the GHG emissions from natural gas relative to coal per equivalent volumes of primary energy. Natural gas has lower well to burner emissions than coal for all points below 1.0 on the vertical axis. The GWP values (x-axis) recommended for methane are 28 and 84 for the 100 and 20 year time-frames, respectively. For both of these values, natural gas is advantageous over coal in terms of GHG emissions as long as methane leakage is less than 3 percent of production (Modified from World Energy Outlook Special Report on Unconventional Gas: Golden Rules for a Golden Age of Gas © OECD/IEA, 2012, fig. 1.5, p. 40).

5.1.2. Analysis of Emissions from Shale Gas Development

Hayhoe et al., (2002) analyzed the GHG emissions of conventional natural gas compared with coal in the context of assessing climatic effects and concluded that replacing coal with gas to generate electricity initially produces higher temperatures (primarily because of reduced sulphate aerosols, and secondarily because of the effects of methane emissions from leakage and venting) but then results in lower global temperatures after 25 years. The study noted considerable uncertainties mainly related to uncertainties in methane emissions (±40 percent) (Hayhoe et al., 2002).

Since 2002, various researchers have attempted to quantify estimated methane emissions better, both from gas production in general and shale gas in particular. GHG emissions are difficult to quantify, involve substantial uncertainties (Venkatesh et al., 2011; Howarth et al., 2012; Logan et al., 2012; Tollefson, 2013b), and for all types of natural gas development include:

- methane and carbon dioxide emissions during drilling and well completion, mostly due to venting and flaring;
- emissions from plays where the gas contains significant proportions of carbon dioxide that has to be removed before the gas can be brought to market;
- methane emissions from fugitive emissions during production, processing, and transport to market; and
- methane emissions from well seeps after abandonment.

Methane can also leak during gas processing and transmission, as discussed by others at length (e.g., Alvarez et al., 2012).

5.1.2.1. Methane Emissions during Drilling and Well Completion

While published estimates of GHG emissions associated with shale gas production vary widely, they generally agree that the most important source of emissions is likely to occur during well completion; that is, after the well has been drilled and before commercial production starts. Once a well is completed, fluids coming back to the surface initially include hydraulic fracture flowback fluids and gas from the producing formation, along with a small amount of granular proppant. Until recently, standard practice was to direct the flowback water into storage and vent or flare the natural gas because equipment was not designed to handle the abrasive mixture of flowback water, sand, and gas. The amount of gas vented or flared over this period depends on the well's production rate, the amount and duration of flowback, the management practices applied, and the nature of the hydraulic fracturing operation. Venting is likely to be more prevalent in the initial stages of a field's development before a gathering pipeline system has been built, and flaring is very common when nitrogen or carbon dioxide are used as the fracturing agent, because the mixture of gases in the flowback water is typically of a non-commercial nature. Reduced emissions completions (RECs), also known as green completions, can capture up to 90 percent of the initial gas flows, reducing the need for flaring (EPA, 2009a).

The gas flared during shale gas well completions is a small percentage of the gas ultimately produced, in all probability less than 0.1 percent. Natural gas flaring is at most 98 percent efficient, so the GHG impact from flaring is about 2.2 times the impact if the gas could be flared at 100 percent efficiency. However, given the far larger impact of consuming all the natural gas produced from a well during its history, any contribution to GHG emissions from flaring completions gas is negligible — far less than one percent of the total (AER, 2011c; B.C. Oil and Gas Commission, 2013e). Flaring, as a result, has declined in both jurisdictions, although recent low natural gas prices represent a disincentive to conservation.

Published estimates of how much methane is released during the completion stage of shale gas development vary by over two orders of magnitude (Howarth et al., 2011; Jiang et al., 2011; Logan et al., 2012; and O'Sullivan & Paltsev, 2012). Some published estimates are admittedly based on a limited number of sites (e.g. Jiang et al., 2011), and others (e.g. Howarth et al., 2011) are controversial because of their magnitude. Burnham et al. (2012) suggested that estimated emissions from shale gas production might in many cases be lower than for conventional gas production. This is because the emissions from liquid unloadings, the removal of liquids that block the flow of natural gas in wet gas wells, are more likely in conventional development as shale gas tends to be dry.

Some experts estimate that methane emissions could be cut by as much as 30 percent through the use of technologies such as "(i) plunger lift systems at new and existing wells during liquids unloading operations; (ii) fugitive methane leak monitoring and repair at new and existing well sites, processing plants, and compressor stations; and (iii) replacing existing high-bleed pneumatic devices with low-bleed equivalents throughout natural gas systems" (Bradbury *et al.*, 2013). Whether such technologies will be adopted — or what sorts of regulatory structures or incentives would be needed to encourage their adoption — is beyond the scope of this chapter.

5.1.2.2. Carbon Dioxide Emissions from Gas Processing in Certain Shale Gas Plays

Natural gas reservoirs may contain small amounts of carbon dioxide that must be separated and disposed of before the gas is brought to market.

Oil recovery or injecting it into coal seams to displace methane for recovery, solutions that are not judged to be as beneficial as CCS from the perspective of preventing climate change.

5.1.2.3. Methane Emissions from Leakages during Production and Processing

There is no science-based evidence that shale gas methane leakage rates are different from those for conventional natural gas development. It therefore appears that the natural gas emissions from exploration and production of unconventional gas may be a significant fraction of total production. This matters because the net GHG impact of coal to gas substitution is highly sensitive to leakage rates (Hayhoe et al., 2002; Wigley, 2011; Alvarez et al., 2012).

Greenhouse Gases and Other Air Emissions 93

5.1.2.4. Methane Emissions from Well Leakage and After Abandonment

When gas production from a well is no longer economical, the well is plugged and abandoned.[12] The well is plugged within the casing with cement to isolate different producing zones, prevent emissions, and protect groundwater. Nevertheless, as a result of the gradual deterioration of materials or inadequate initial well construction, many abandoned wells leak either through the wellbore (surface-casing vent flow) or around it (gas migration outside casing) (Dusseault et al., 2000; Watson & Bachu, 2008). The source of these leaks is often not theoriginal production formation but an intermediate gas-bearing formation that had not been in production because it was either not known to exist or not of commercial value (Muehlenbachs, 2012b).

The proportion of abandoned wells that leak is difficult to estimate.

Cement failure is the main cause of methane leakage (Watson & Bachu, 2009). Over time, the cement (and/or casing) tends to deteriorate, allowing buoyant gas to leak along the annulus between the production casing and the formation (Dusseault et al., 2000). This deterioration explains why older wells leak more. In fact, Mueller and Eid (2006) warn that the pressure testing that occurs soon after the cementation of the surface casing may cause severe tangential stresses on the cement sheath, causing it to fail. It is therefore important to monitor the integrity of the well condition after plugging and abandonment.

5.1.2.5. Well-to-Burner Analysis

Some studies have estimated the total well-to-burner GHG emissions for shale gas, including production, processing, transport, and consumption. Figure 5.2 presents these results graphically, comparing the life-cycle GHG emissions estimated for the Barnett with the range of estimates for the well-to-burner emissions associated with coal, conventional natural gas, and

---

[12] Regulatory authorities allow production from a gas well to be suspended temporarily but place time limits on such suspensions.

unconventional natural gas (mostly shale) after methodological harmonization. Harmonization is a process that compares different studies by ensuring they use a consistent set of included processes and metrics, and variability is reduced by setting primary energy resource characteristics and/or key performance parameters to consistent values based on some modern reference system (Heath & Mann, 2012; and Whitaker et al., 2012). For comparison purposes, Figure 5.3 shows well-to-burner emissions for electricity generation using different energy sources.

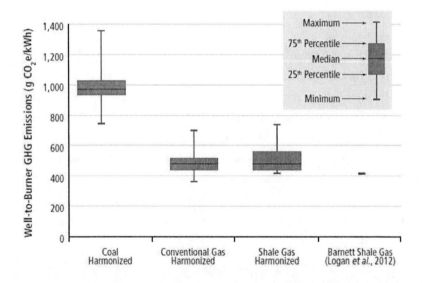

Figure 5.2. Well-to-Burner Estimates for the GHG Emissions Associated with Electricity Generation from Coal, Conventional Gas, Unconventional Gas, and Barnett Shale Gas After Methodological Harmonization. The range of estimates for well-to-burner emissions in grams of carbon dioxide emissions per kilowatt-hour for coal, conventional natural gas, and unconventional gas, based on several different studies (see Logan et al., 2012). The graph also shows the value of the well-to-burner emissions estimated by Logan et al., (2012) for shale gas extracted from the Barnett Shale. The authors found that these emissions were approximately the same as those from conventional gas and therefore roughly half those of coal. The conventional, unconventional, and Barnett Shale values are for electricity generation in a natural gas combined-cycle turbine.

Greenhouse Gases and Other Air Emissions

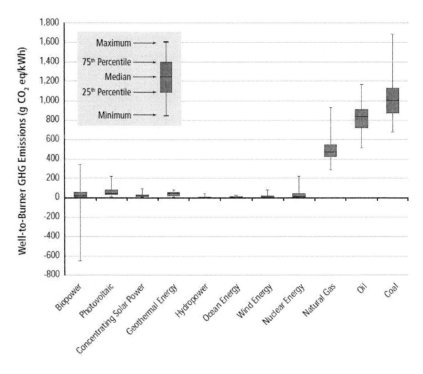

Figure 5.3. Range of well-to-burner emissions for electricity generated from different sources. The life cycle (well-to-burner) GHG emissions for electricity generation using different sources. Coal emits the greatest amount of GHGs, while renewable sources have very low GHG emissions. The emissions from natural gas are less than those of coal, but greater than those from renewable energy sources and nuclear.

Several articles present results of upstream and midstream emissions that are consistent with those of Logan et al., (2012) (e.g., Burnham et al., 2012; Cathles et al., 2012) whereas others reach different conclusions (Howarth et al., 2011; Hultman et al., 2011). In 2011, Howarth et al., (2011) published a study that argued that upstream leakage rates of unconventional gas are so significant that the net GHG impact of unconventional gas may be worse than coal, particularly when viewed on the 20-year time frame (rather than the 100-year time frame). They defended this time-frame for analysis, arguing that the 20-year time frame is appropriate given the need to avoid climate tipping points and that the analysis of electricity generation sources is a relevant but incomplete framework (Howarth et al., 2011). Others have questioned the conclusions of Howarth et al., (Cathles et al.,

2012) or produced estimates that are significantly lower (Jiang et al., 2011; Burnham et al., 2012; Weber & Clavin, 2012; O'Sullivan & Paltsev (2012). Wigley (2011) comes to a comparable conclusion, arguing that for completion methane leakage rates greater than two percent, methane leakage offsets the beneficial effects of carbon dioxide reduction that accompanies the transition from coal to gas. Scientific studies of leakage rates are on-going, and the scientific evidence remains in flux. The general trend of recent studies suggests that the earlier estimates by Howarth et al., (2011) might be too high, but whether or not actual rates are low enough to preserve the overall GHG benefits of shale gas over coal remains a subject of study.

The different conclusions reached in these publications demonstrate the difficulty in estimating emissions precisely, which helps to explain why evaluating the environmental impact of shale gas with respect to GHG emissions and climate change continues to be debated. Research to address this uncertainty is underway. However, given the technical difficulty of measuring methane leakage accurately and the number of considerations that contribute to the various conclusions that experts have reached to date, it seems unlikely that the uncertainties will be resolved to general satisfaction in the immediate future.

5.1.3. Knowledge Gaps

As discussed above, the primary knowledge gap related to the impact of GHG emissions associated with shale gas development stems from the uncertainty in estimating the total methane emissions themselves. In particular, better measurements of well leakage, including measurements to assess gas leakage outside the well casing, and field-proven estimates of the number of leaking wells, are necessary. There are also differences in scientific opinion regarding the appropriate values to use in evaluating the warming potential of methane, the impact of atmospheric aerosols produced by burning coal, and the type of combustion technology applied. Most well-to-burner analyses focus on the use of gas to replace coal in electricity

Greenhouse Gases and Other Air Emissions 97

generation, where a substantial portion of the benefit that accrues arises from the greater efficiency of combined-cycle gas turbines.

On the other hand, shale gas could also potentially displace low-carbon forms of electricity generation, which would increase GHG emissions. The net effect will depend on a number of factors, such as the future prices of various fuels, and energy and climate protection policy decisions made in countries importing liquid natural gas.

5.2. OTHER AIR EMISSIONS

5.2.1. Chemical Emissions

The air emissions attributable to shale gas development typically come from the same sources (e.g., drilling rigs, truck engines, gas compressors, holding ponds, vents, and flares) as those associated with conventional gas production and, indeed, other forms of mining and industrial activity. The main difference is that these sources may be produced more intensively in shale gas development (due to longer drilling times, more trucks being used, more powerful pumps, and bigger holding ponds) because of the added effort required to extract gas from shale.

Air emissions from these sources include NOx and SOx, particulate matter, BTEX, and other hazardous air pollutants. VOCs and other pollutants associated with natural gas and fracturing fluids can enter the air from both wells and activities associated with flowback water (separators, pits, or tanks) (Gilman et al., 2013). When combined with NOx and carbon monoxide, VOCs can act as precursor chemicals to the creation of ground-level ozone, a known cause of respiratory disease (EPA, 2012b; McKenzie et al., 2012). As the individual sources of pollutants may be widely distributed, it can be difficult to estimate total emissions.

The concerns associated with some of the air emissions from shale gas development. The impact of these emissions depends on their magnitude and, in the case of health effects, the length of exposure.

As development grows in scale, mobile diesel engines can be replaced with fixed electric or compressed natural gas (CNG) engines, thereby reducing air emissions.

Table 5.2. Air Emissions and Concerns Associated with Shale Gas Development

Substance	Source	Concern
NOx, SOx, VOCs	Diesel engines, natural gas compressors, fluid evaporation	Ozone precursors (smog) Impacts on human health (e.g., lung disease)
BTEX and other HAP	Venting, fugitive emissions, flaring, fluid evaporation	Potential impacts on nervous system
Particulates (PM 2.5)	Diesel engines, flaring	Lung diseases (air quality)
Methane	Venting, fugitive emissions	GHG emissions (climate change)
Carbon dioxide	Diesel aggregates, flaring, fugitive emissions	GHG emissions (climate change)

Data Source: McKenzie et al., 2012. PM 2.5: Particulate matter smaller than 2.5 microns; BTEX: Benzene, toluene, ethylbenzene, and xylene; HAP: Hazardous Air Pollutants[13].

5.2.2. Knowledge Gaps

Information is needed on the location of future wells, the planned infrastructure, and the anticipated scale of development to assess the impact of air emissions. In addition, baseline observations of air quality are lacking in several regions where development has taken place or may take place. Contaminant dispersion models are also needed to understand the impact of the air emissions associated with the industry.

---

[13] In 1990, the U.S. Clean Air Act identified 188 specific pollutants and chemical groups as hazardous air pollutants (HAP). The list has been modified over time. See GoodGuide (2011) for a list of HAP.

5.3. CONCLUSION

Detailed studies of shale gas environmental benefits and risks in terms of GHG emissions remain to be undertaken. It is recognized that regulations and accepted practices are somewhat more stringent (e.g. fully cemented production casing, less venting, no surface lagoons for flow-back fluids) than in the where much of the data have been generated, but much uncertainty remains. Reducing this uncertainty is important in evaluating the environmental impacts of shale gas development, and data can clearly point the way for improvements at all stages of shale gas development and consumption.

Chapter 6

LAND AND SEISMIC IMPACTS

Chapters 4 and 5 addressed the environmental implications of shale gas extraction as they relate to water, in terms of quality and quantity, and air, in terms of GHG and other emissions. The impact that large-scale shale gas development has on land resources, notably its potential cumulative effects and their environmental impacts, and the potential risks associated with stimulated seismicity, are of concern. These issues are discussed below.

6.1. LAND IMPACTS

The development of energy resources is one of the primary drivers of land and terrestrial ecosystem changes (Northrup & Wittemyer, 2013). Conventional oil and gas development, for example, involves an extensive infrastructure that includes seismic lines, well pads, work camps, waste handling, compressor or pumping stations, processing plants, gathering lines, and transmission pipelines. Creating this infrastructure requires that land be cleared, gravel be quarried, and roads and bridges be built, often to access remote areas.

As discussed in Chapter 3, shale gas development involves the same mix of construction and industrial activities as conventional gas development but at a higher intensity because: (i) the resource covers large geographical areas; (ii) production declines quickly requiring a large number of wells to be drilled to keep production stable; and (iii) individual shale gas wells need to be spaced more tightly together to drain the reservoir efficiently due to the rock's low permeability. In terms of land impacts, however, it is the pad size and its spacing (as opposed to well spacing) that is most significant. Having multiple wells on a single pad is environmentally preferable.

Development may take place over several decades from first exploration to the completion of land reclamation activities. While the pace of activities will peak at the drilling and completion stage, and decline sharply during production, land impacts will be seen and felt for a very long time.

The nature and significance of these impacts will vary depending on regional differences in physical features, scale of development, and population density as well as the scale, complexity, and intensity of existing land uses. Nevertheless, it is possible to characterize the principal sources of land impacts from shale gas development — that is, the activities that could lead to important changes to the land — and their generic environmental effects. It is also important to also consider how these activities and their effects may be borne out in the different shale gas plays.

6.1.1. Sources of Impact

The main shale gas development impacts on land and ecosystems arise from the (i) construction and operation of well pads; (ii) the construction of access roads and increased vehicular traffic; and (iii) the construction of other infrastructure, such as for water storage. Shale gas development also involves the construction of gas-processing plants and transmission pipelines; these fall outside the scope of this chapter.

6.1.1.1. Well Pads

Though the size of individual pads and their construction technique can vary, shale gas well pads are often larger than conventional gas pads —

taking up, on average, some 3 hectares of land compared to 1.9 for conventional drilling (Broomfield, 2012). They may hold several wells and need to accommodate more equipment and trucks during the drilling and completion stages. The land above a productive shale gas zone could be disturbed during drilling during production.

As the technology to drill longer horizontal laterals improves, it is becoming possible to place more wells on an individual pad, reducing the industry's footprint, including the number of roads to be built, the number of sites storing hazardous chemicals, and conflicts with other land uses. For example, a pad with four wells in each direction (for a total of eight, see Figure 6.1), with horizontal laterals 1,500 metres long spaced 250 metres apart could drain a 3 square kilometre area. A pad with eight wells in each direction could drain an area twice as large. Figure 6.1 shows a multi-well pad.

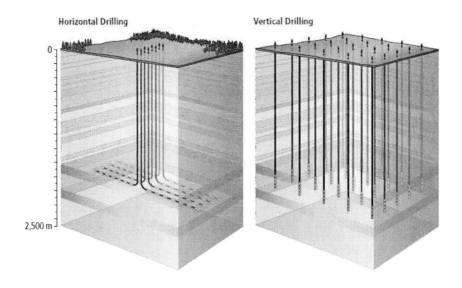

Figure 6.1. A Multi-Well Pad versus a Cluster of Single Well Pads. Schematic illustration of horizontal shale gas wells and vertical wells. Through the use of multiple wells on a single pad and longer laterals, a greater area is covered by each well pad, reducing the needed pad density.

Figure 6.2. Area with a Multi-Well Pad and operational shale gas wells Schematic illustration of horizontal shale gas wells and vertical wells. Through the use of multiple wells on a single pad and longer laterals, a greater area is covered by each well pad, reducing the needed pad density (modified from Nexon Energy ULC).

If the area that is being developed has more than one zone at depth, the same pad could be used to drill another series of wells to drain a secondary layer. Figure 6.2 shows an area with a Multi-Well Pad and operational shale gas wells Schematic illustration of horizontal shale gas wells and vertical wells. Through the use of multiple wells on a single pad and longer laterals, a greater area is covered by each well pad, reducing the needed pad density.

6.1.1.2. Access Roads and Vehicular Traffic

In areas with little existing infrastructure, such as in parts of, road construction to service drilling sites can be an extensive undertaking. In northern areas of the province, some of these roads may be temporary winter roads, but in most areas of shale gas potential, they are likely to be permanent and open up new territories to development pressures (e.g., logging, hunting, fishing, and cottaging). Roads may already exist but are seldom built to carry the heavy trucks the industry requires and many will require upgrading.

6.1.1.3. Supporting Infrastructure

Depending on local conditions, supporting infrastructure, including gravel pits, staging areas, supply yards, utility corridors for water and gas pipelines or electrical lines, compressor stations, and water storage facilities

Land and Seismic Impacts 105

may need to be built or expanded. In some areas, this associated infrastructure may consume more land than the well pads themselves (Johnson, 2010). The storage of water for hydraulic fracturing can also require the construction of impoundments or pits.

An additional infrastructure feature of shale gas development is the need for propant. The spread of hydraulic fracturing technology oil and gas industry has led to a sharp increase in demand for high-quality silica to use as proppant in well stimulations. A 20,000 cubic metre hydraulic fracturing treatment may use one and a half million kilograms of proppants (King, 2012).

Companies that use proppant sand look for specific characteristics (e.g., high silica content, uniform size, round shape, high crush resistance) to hold open induced fractures in a given geological formation. Consequently, several varieties of proppant sand exist, each with its own attributes. Sources of proppant sand could include both consolidated (weakly cemented sandstone) and unconsolidated (Quaternary) sources.

The environmental impacts of mining proppant sand are similar to those of large-scale quarrying operations (e.g., dust, noise, and scarring of the land). In addition, the fine dust associated with mining and transporting fine sand can adversely affect health if inhaled (McLeod, 2011) (*viz.*, silicosis; see Chapter 7).

6.1.2. Forest and Wildlife Impacts

When shale gas development occurs in forested areas, land must be cleared to allow the necessary well pads, roads, and infrastructure to be built. Depending on well density and other factors, large-scale clearing may be necessary with significant consequences for ecosystem integrity and wildlife.

Figure 6.3 shows an aerial view of shale gas development and illustrates the spatial impact that the well pads and supporting infrastructure (e.g., roads) have had on that forested land area.

The activities related to shale gas development described above can be expected to affect forest ecosystems by fragmenting them (i.e., intersecting and sub-dividing them with roads, pipelines, and other works) and creating edge effects (i.e., transition zones between disturbed and undisturbed habitats) (Johnson, 2010). Both phenomena disrupt ecosystem structure and function and change resource availability for wildlife and the physical environment. Large contiguous forests have many traits not shared by collections of small patches equal to the same size. These include being more resistant to the invasive species, providing more habitat for forest plants and animals, suffering less damage from storms, and providing more ecosystem services, such as water filtration (Johnson, 2010).

Figure 6.3. An aerial view of landscape disturbance caused by shale gas development in a Shale Gas Infrastructure (courtesy of Hayley Dunning).

"Changes in land use and land cover affect the ability of ecosystems to provide essential ecological goods and services, which, in turn, affect the economic, public health, and social benefits that these ecosystems provide" (Slonecker, 2012).

There are few studies on the impacts of unconventional oil and gas development on wildlife other than on certain types of birds and large mammals (Northrup & Wittemyer, 2013). On environmental effects of resource roads in construction and use are mostly detrimental in terms of the environment, although they can have clear social and economic benefits (as

Land and Seismic Impacts 107

reviewed by Daigle, 2010). Specific impacts will vary depending on a number of factors (e.g., ecosystem characteristics, road density, and extent of road use), but can include:

- loss of biomass productivity from displaced and compacted soil;
- increased erosion leading to sediment and nutrient delivery to streams and wetlands;[14]
- changed hydrological regimes (e.g., altered streamflow), increased sedimentation, and loss of aquatic habitat;
- restricted fish passage due to road infrastructure (e.g., bridges) and increased fish mortality;
- increased wildlife mortality and injuries because of expanded hunting, fishing, and trapping;
- loss of species, habitat, and vegetation;
- fragmented wildlife habitat and disrupted migration patterns;
- increased human disturbance of wildlife; and
- contaminant emissions such as road salt, oil, and gasoline.

(as reviewed by Daigle, 2010)

The transport of equipment may also bring invasive species to the project site. Depending on how they spread, new species can disturb local ecosystems and adversely affect the local economy (e.g., agriculture) (Daigle, 2010). Finally, there is the issue of what happens to a road after a company abandons a well site.

6.1.3. Regional Considerations

By definition, the potential impacts on land from large-scale shale gas development will depend on the geographical setting in which they occur. In the absence of information on the location, pace, and scale of

---

[14] This is also an issue for the construction of well pads. Notably, most industry regulatory violations in Pennsylvania between 2008 and 2011 involved improper erosion and sedimentation plans (Staaf, 2012).

development, the following provides short contextual information about the regions in which shale gas development may take place.

6.1.4. Land Reclamation and Remediation

As described above, large-scale shale gas development exerts a large footprint on the land. The land area covered by production wells will be less than that needed for development (on average 25 percent less for a multi-well pad) (NYSDEC, 2011), and some land used during well drilling and completion can be reclaimed. According to CAPP (2013b), fully reclaiming a well site takes about five years and includes capping the well, removing equipment, carrying out needed remediation, and replanting vegetation. Full restoration of sites may not be possible in many cases, notably in areas of high agricultural, cultural, or natural value. While careful siting of wells can reduce local impacts, large-scale development could result in a significant loss or fragmentation of natural processes and/or existing land uses (e.g., recreational, agricultural, or natural).

The degree of land reclamation that may be needed from shale gas development, generations into the future, is uncertain and thus consideration should be given to the risk and financial liability that arises from such uncertainty.

6.1.5. Knowledge Gaps

The land impacts from shale gas production are not well-defined though many can be inferred from to complete baselines studies to be able to assess regional impacts (e.g., species at risk, critical wildlife habitats, surface hydrological regimes), the bigger challenges may be institutional and methodological. Shale gas development implies a large number of activities by multiple operators, extending over a large geographical area and a long period of time. With government agencies, at all levels overseeing those aspects of the development within their mandate, there is a risk that jurisdictional fragmentation will result in a piecemeal assessment of environmental effects that does not reveal the full implications of shale gas development. In addition, the methodologies for studying cumulative effects are not well-developed and will require more effective implementation of strategic impact assessment processes. Moreover, the need for post-operational cumulative effects monitoring should not be underestimated. Even with full compliance, unforeseen cumulative consequences of development may only be detected and addressed through post-operational monitoring and adaptive management.

6.2. STIMULATED SEISMICITY

The mechanisms by which fluid injection can stimulate seismicity (i.e., cause earthquakes) are well understood. The fact that earthquakes can be triggered by fluid injection was recognized as early as the 1920s. Thus, the principal physical mechanisms responsible for injection-related seismicity have been known for almost 50 years: increasing pore pressure at depth reduces the effective normal stress acting perpendicular to pre-existing faults at depth. If faults affected by increased pore pressure are potentially active, the elastic strain energy already stored in the rock can be released by fault slip and earthquakes.

The main determinants of stimulated seismicity are the:

- pressure change at depth;
- interaction between the pressure change and pre-existing faults;
- volume of injected fluid (larger volumes generate larger pressures acting over larger areas); and
- rate of injection in the case of wastewater injection (as more rapid injection generates higher pressure).

In discussing the risk of unintended seismicity caused by hydraulic fracturing, it is useful to differentiate between two types of stimulated seismicity (McGarr et al., 2002 as referenced in NRC, 2012):

- *Induced seismicity* – is the development, due to human activity, of new fractures that change the distribution of stress in geological formations (e.g., through pumping fluids in or out of a formation). Induced seismicity has been studied in several other situations (e.g., carbon dioxide injection, water injection, geothermal applications, mining) for many years. Induced seismic events are usually small enough that most individuals will not feel them at the surface.
- *Triggered seismicity* – occurs when human activity causes seismic events that might otherwise have happened in the future. Triggered seismicity along major fault lines (>10 km long, depending on depth) can cause seismic events of sufficient magnitude to damage buildings and disrupt public life. Activities in geologically active areas need to be carefully monitored to minimize the risk of triggering higher magnitude seismic events.

There are two points in the shale gas development process at which seismic events could be induced or triggered: during hydraulic fracturing as a well is being stimulated, and during wastewater injection as an operator disposes of liquid wastes. These possibilities are discussed below.

6.2.1. Hydraulic Fracturing

The injection of fluids at high pressures to infiltrate existing fractures and rock pores triggers thousands of micro-seismic events as shear and tensile fractures are created that change the permeability of the shale, allowing gas to flow to the wellbore. The reach of the fractures generated is relatively small (up to 300 metres) compared to the depth of the well (often 2 to 3 kilometres), and the injection process itself lasts only a few hours. The operators then reduce the stress by allowing the injected fluids to flow back

Land and Seismic Impacts   111

to the surface. While most of the induced seismicity occurs during the fracturing, the stress pattern in the formation is affected during the entire production period and may take several years after abandonment to reach a new equilibrium.

Cases have been documented in which hydraulic fracturing was identified as the cause of unintended minor seismic events. Such unintended seismicity associated with shale gas development is rare. Seismic events of similar magnitude, attributable to the operations of the oil and gas industry (mostly conventional oil and gas production), have also been noted in several areas. While such seismic events are too small to cause property damage, a few have been felt at the Earth's surface and have given rise to public concerns about the safety of hydraulic fracturing.

Box 6.1. Recommendations on Seismicity

After an investigation of observed seismicity recommended the following:

i.   Improve the accuracy of the Seismograph Network.
ii.   Perform geological and seismic assessments to identify pre-existing faulting.
iii.   Establish induced seismicity monitoring and reporting procedures and requirements.
iv.   Station ground motion sensors near selected communities to quantify risk from ground motion.
v.   Study the deployment of a portable dense seismograph array to selected locations where induced seismicity is anticipated or has occurred.
vi.   Require the submission of micro-seismic reports to monitor hydraulic fracturing for containment of micro fracturing and to identify existing faults.
vii.   Study the relationship between hydraulic fracturing parameters and seismicity

---

The industry maintains databases on the seismicity induced by their operations during hydraulic fracturing. However, few companies monitor their activities for seismic events once satisfied that they understand the properties of the shale and the propagation of fractures. Operators and service companies state that it is in their interest to ensure the reliability of these data to maximize the efficiency of their fracturing operations. Data are shared with regulators, although for commercial reasons they are rarely publicly available.

6.2.2. Wastewater Injection

The standard practice in the oil and gas industry is to dispose of contaminated fluids by injecting them underground. Injection wells are sometimes shallower than production wells but still much deeper than freshwater aquifers. Injection typically involves greater fluid volumes per well than is the case for hydraulic fracturing operations, albeit at lower pressures. Waste fluids are injected into porous formations specifically targeted to accommodate large volumes of fluid (often-depleted oil and gas reservoirs). Deep injection disposal of waste fluids from hydraulic fracturing and other industry activities is regulated wherever it takes place and must avoid the fracturing of rock. The relationship between induced earthquakes and energy industries also examined the issue of wastewater injection and concluded that "injection for disposal of waste water derived from energy technologies into the sub-surface does pose some risk for induced seismicity, but very few events have been documented over the past several decades relative to the large number of disposal wells in operation." Furthermore, the report stated that the majority of wastewater wells pose no risk in terms of induced (or triggered) seismicity, but that the long-term seismic effects resulting from of a large increase in the number of these disposal wells are not known. A more recent study, however, argues that the recent unusual increase in earthquake activity in the continental interior may be a delayed reaction to many years of wastewater injection in Oklahoma (Keranen *et al.*, 2013).

Land and Seismic Impacts 113

Traffic light monitoring is a part of a management approach for monitoring and acting on the magnitude of stimulated seismicity during hydraulic fracturing if pre-determined thresholds are reached (Zoback, 2012). Other steps can also be taken to reduce the probability of triggering seismicity when injecting waste. These include:

- Developing an understanding of the stress state, pore pressure, location, orientation, and size of pre-existing faults to avoid injecting into active faults.
- In areas of concern, carrying out real-time seismic monitoring.
- If there are triggered earthquakes, utilizing the information above to develop a mechanistic understanding of the triggered seismicity.
- Applying *If...then...* rules, agreed upon ahead of time with the appropriate regulatory agencies (e.g., if an earthquake of a certain magnitude and characteristics occurs then injection should immediately cease).

6.2.3. Knowledge Gaps

The knowledge gaps that need to be filled to manage seismic events associated with hydraulic fracturing (Box 6.1). Similar gaps are likely to exist in other regions where shale gas development is possible.

While the science of how fluid injection can cause seismic events is well understood, and risks can be minimized by being proactive, whether wastewater injection can be safely carried out in all regions is unknown.

More information on the potential for geological formations in these provinces to receive large volumes of injected fluids without over-pressurizing reservoirs is needed to determine whether this waste disposal option is possible. If it is not, shale gas developers who use slickwater technology will need to find alternative water disposal methods prior to development.

Chapter 7

HUMAN HEALTH

During its deliberations, a broad definition of the term *environment*, referring not only to the natural environment — the physical and biological systems that support us — but also the social environment in which we live. While humans shape the environment, they are also part of it. The development of shale gas will affect not only the air, water, and land of the regions where production may take place but also the health of the people who live there.

The effects of shale gas development on human health have not received much scientific and government attention despite often being cited as an issue of public concern (Goldstein et al., 2012, 2013). Significant gaps exist in our understanding. Where development has taken place, public debate has tended to focus on the physical risks of exposure to various chemicals (OCMOH, 2012). However, factors related to the physical, social, and economic environments in which people live (e.g., community disruption) can also adversely affect health.

The risks to human health that shale gas development poses are in many ways similar to those of the conventional oil and gas industry (e.g., air emissions from heavy machinery) and indeed those of many large-scale resource developments (e.g., boomtown phenomenon). Nevertheless, some risks — water pollution from hydraulic fracturing chemicals and flowback water, for example — are specific to unconventional oil and gas development.

7.1. DEFINITION AND RISK FACTORS

To understand the potential health impacts of shale gas development, it is necessary to have a clear definition of human health and an understanding of the factors that influence it. The World Health Organization defines health as both "a state of complete physical, mental and social well-being and not merely the absence of disease or infirmity" and "the extent to which an individual or a group is able, on the one hand, to realize aspirations and to satisfy needs, and on the other, to change or cope with the environment" (WHO, 1946, 2009). Health encompasses both the physical well-being of the individual and incorporates the social, emotional, spiritual, and cultural well-being of the whole community.

Figure 7.1. Determinants of Health (Modified from Health and the Environment: Critical Pathways. Health, 2002. The range of factors that influence human health. These factors include social, economic, and environmental determinants, in addition to genetic ones (Reproduced from the Minister of Health, 2014).

Human Health 117

Human health is influenced by various determinants, including environmental (e.g., water quality), socio-economic (e.g., income, employment) and cultural (e.g., attachment to specific geographical locations). The identified 12 such determinants of health (ACPH, 1994, 1999) (Figure 7.1).

The occurrence and significance of impacts of shale gas developments on physiological human health will depend on a number of local factors, such as the following:

- Population density/proximity of development to settlements: the risks of exposure will depend on the makeup of the population (e.g., demographic characteristics, the baseline health status). In areas of low density, individuals who rely on the harvesting of country food (wildlife, fish, produce) may also be adversely affected.
- Some groups, including oil and gas industry workers, children, hypersensitive individuals, and in some regions.
- Ambient environmental conditions such as existing sources of background pollution.
- Geology determines the chemistry of the fracturing fluids to be used and the substances returning with the flowback water (e.g., NORM, barium, salt) as well as the ability to dispose of flowback water by deep injection.
- Legal/regulatory framework: shale gas development is primarily regulated at the provincial or state level. Different jurisdictions impose different requirements to protect human health, public safety, or to regulate social impacts.
- Workforce training and oversight.
- Frequency and intensity of development.

Links have been suggested between oil and gas development and local community health impacts (Witter et al., 2008; Steinzor et al., 2013). While most studies focus on the undesirable effects that shale gas development may have on people living near wells or working in the industry, some have argued that shale gas development may have a beneficial impact on human health. To the extent that it displaces coal for electricity generation, gas combustion emits fewer harmful air pollutants, such as particulates, sulfur oxides, nitrogen oxides and mercury, and has less of a global climate effect. Long-term exposure to airborne particulates has been documented to be associated with mortality (Crouse et al., 2012).

The paucity of data and evidence of causal links in this area results from the lack of baseline studies, inadequate monitoring, and in some cases non-disclosure agreements may make it challenging to document incidents of contamination (Bamberger & Oswald, 2012). The very nature of non-disclosure agreements makes it impossible to know how many there are and what matters they cover. According to an article in Bloomberg, a financial news service, "the [non-disclosure agreements] keeps data from regulators, policymakers, the news media and health researchers, and makes it difficult to challenge the industry's claim that fracking has never tainted anyone's water" (Efstathiou & Drajem, 2013).

Some researchers have therefore studied companion animals as they often share the same exposures as humans indicates that animal health can serve as a sentinel for human health. Impact of gas industry operations (both conventional and shale gas) on farm and companion animals, for example reproductive problems, and other health effects resulting from inappropriate industry practices, spills, accidents, compressor malfunction, and gas flaring (Bamberger & Oswald, 2012).

The rest of this chapter describes what is known about the main health stressors associated with the shale gas industry and their implications for human health.

7.1.1. Occupational Health

Working on a drilling site exposes workers to a number of hazards, including accidents involving heavy machinery and exposure to chemicals. Hydraulic fracturing may add new risks through exposures to an added suite of chemicals and physical agents on a work site (Cottle & Guidotti, 1990). An increasingly recognized health risk to workers is the inhalation of silica,

Human Health

119

used as a proppant in hydraulic fracturing which can cause silicosis, lung cancer, and other diseases (NIOSH, 2002). Workers may also be at risk because of exposure to naturally occurring radioactive material (NORM) brought to the surface through flowback water or drill cuttings. As NORM can accumulate on equipment and machinery, workers can be exposed through contact with skin in addition to inhalation through air and water exposure (Hamlat et al., 2001).

7.1.2. Health Risks from Waste

Hydraulic fracturing uses a large number of chemicals, including some known hazardous substances (e.g., the foaming agent 2-butoxyethanol), and brings many potentially dangerous compounds to the surface, such as hydrocarbons, varying amounts of BTEX compounds, brine, and other naturally occurring geological components (e.g., arsenic, radionuclides). Colborn et al., (2011) evaluated the hazards of 353 chemicals that the natural gas industry uses in its operations, a subset of the 632 chemicals in 944 products identified. Of note, for over 400 of the products identified, less than one percent of the total chemical composition was available, leading to a substantial degree of uncertainty concerning the risk that these products may pose to human health (Colborn et al., 2011). The author carried out literature searches to determine the potential health impacts of the chemicals identified and found that "75 percent can affect sensory organs and the respiratory and gastrointestinal systems; 40 to 50 percent can affect the nervous, immune, and cardiovascular systems as well as the kidneys; 37 percent can affect the endocrine system; and 25 percent can cause cancer and mutations." It should be noted that these risks may stem from ingestion or direct dermal exposure and these chemicals are often used in very low concentrations (Colborn et al., 2011).

One of the issues of greatest toxicological concern is that of the potential impact of untested mixtures of chemicals (Goldstein et al., 2013).

Pathways for exposure include contamination of water by spills, leaks, or unintended underground communication between the production zone and shallow aquifers and to air as a result of evaporation from condensate tanks and flowback storage. Such contamination could affect individuals directly (e.g., where well water becomes contaminated) or indirectly through the food chain (Fraser Basin Council, 2012).

Methane is also a contaminant of concern for drinking water (as discussed in Chapter 4), while methane is not generally considered a health hazard when ingested, at high concentrations it can cause asphyxiation if inhaled in confined spaces (as can most gases) (Cooley & Donnelly, 2012). It also poses fire and explosion hazards at elevated concentrations.

7.1.3. Health Risks from Air Pollution

The oil and gas industry is also a source of air contaminants (as discussed in Chapter 5). Table 7.1 lists the potential health effects of these air pollutants. It is important to note that a specific health effect and its extent will depend on a variety of factors such as the type and length of exposure to a contaminant as well as the health status and lifestyle of the exposed individual (CDPHE, 2010). The potential toxicity of a contaminant does not necessarily reflect the health effect. (Kemball-Cook et al., 2010; Litovitz et al., 2013). Such emissions have led a European report to conclude, based on independent but non-peer reviewed sources, that large-scale shale gas development could significantly increase ozone levels in some areas, leading to a potentially high risk of adverse effects on respiratory health (Broomfield, 2012).

Exposure to air pollutants associated with shale gas development may lead to a small increase in the risk of cancer and other diseases such as neurological and respiratory effects for people living in close proximity to a well. The primary contributor to cumulative cancer risk was benzene, whereas other airborne VOCs such as trimethylbenzenes, xylenes, and aliphatic hydrocarbons were the primary contributors to the subchronic non-cancer hazard index (McKenzie et al., 2012). While benzene emissions resulting from activities did not exceed short-term exposure standards, they could pose a risk to human health if their levels were representative of long-term ambient conditions (Ethridge, 2010).

Table 7.1. Air Pollutants Associated with Shale Gas Development and Their Potential

Substance	Potential Health Effects
Particulate Matter (PM)	• non-fatal heart attacks • irregular heartbeat • aggravated asthma • reduced lung function • increased respiratory symptoms (e.g., coughing, difficulty breathing) • premature death in people with heart or lung disease
Nitrogen Oxides (NOx)	• irritated respiratory system • aggravated asthma, bronchitis, or existing heart disease
Carbon Monoxide (CO)	• exacerbation of cardiovascular disease • behavioural impairment • reduced birth weight • increased daily mortality rate
Volatile Organic Compounds (VOCs) (e.g., BTEX)	• carcinogen (some VOCs) • leukemia and other blood disorders (benzene) • birth defects (some VOCs) • eye, nose, and throat irritation (some VOCs) • adverse nervous systems effects
Methane (CH_4)	• asphyxiation in confined spaces
Ground Level Ozone (O_3) (Smog)	• reduced lung function • aggravated asthma or bronchitis • permanent lung damage

Data Source: Fierro et al., 2001; McKenzie et al., 2012.

In addition, exposure to NORM can occur through drilling mud or flowback water brought back to the surface (Hamlat et al., 2001). Run-off or evaporation from flowback pits, for example, may contaminate agricultural lands, and hence, feed crops with potentially harmful radioisotopes (Rich & Crosby, 2013). Potential health risks from exposure to flowback pit contents through this mechanism have not been established (see references in Rich & Crosby, 2013). Increased exposure to radon within homes that use natural gas from the Shale can also been raised as a possible concern (Resnikoff, 2012), although several studies of water and oil and gas equipment strongly dispute that this is a risk (Johnson, 2010).

7.1.4. Psychosocial Impacts

Impacts on psychological well-being are important to consider because of their effect on human health (*viz.*, social environments and coping skills as two determinants of health). Studies suggest that individuals who believe they have been affected by shale gas development exhibit various symptoms, including increased fatigue, nasal and throat irritation, sinus problems, eye irritation, shortness of breath, joint pain, severe headaches, and sleep disturbance (Steinzor et al., 2013). Public health can also be negatively affected by the cumulative impact of social stressors including socio-economic changes, social change (e.g., increase in crime), and change in the nature of communities (Korfmacher et al., 2013).

Lack of transparency, conflicting messages, and the perception that industry or authorities are not telling the truth can create or augment concerns about one's quality of life or well-being, and contribute to feelings of anxiety about the potential health, environmental, or community impacts (Fraser Basin Council, 2012). Uncertainty about whether changes in quality of life will be temporary or not, can feed anxiety about the future (Perry, 2013). A particular public concern related to shale gas development has been secrecy about the chemical and physical agents that are added to the hydraulic fluids or are brought to the surface (BAPE, 2011b; Committee of Energy and Commerce, 2011).

In addition to direct impacts, shale gas development activities can exert indirect impacts on human health. Changes to ecosystem health (e.g., arrival

Human Health 123

of invasive species) or to the public health infrastructure (e.g., shortage of medical staff because of a booming population) can also affect the health of a community indirectly.

7.1.5. Community Disruption and Quality of Life Issues

Shale gas development can result in rapid population increases, particularly in isolated rural areas. Where there are few economic opportunities, the growth of a new industry can diversify the local economy, create jobs, and slow down or reverse rural out-migration. Such economic development has the potential to improve population health.

On the other hand, such gains may also imply trade-offs if they compromise economic activities such as tourism or fishing. The influx of temporary workers can disrupt existing community patterns by increasing social pathologies (e.g., crime rates, substance abuse, sexually transmitted infections) (Fraser Basin Council, 2012; OCMOH, 2012) and contribute to local inflation. Inflation often disproportionately affects residents not directly associated with the industry (Jacquet, 2009; OCMOH, 2012), thereby increasing income inequality. In addition, communities are often unable to meet the demands of the growing populations due to the difficulty of forecasting the extent, timing, and location of development, which results in inadequate housing, policing, emergency preparedness, and health care and social services (Eligon, 2013). Symptoms of this phenomenon, known as the *boomtown effect*, have been documented in many communities in both the United States and Canada (Jacquet, 2009, 2013; OCMOH, 2012). This effect can have a negative cumulative impact on public health (Goldstein et al., 2013; Korfmacher et al., 2013).

Under common law, property owners or renters are entitled to the quiet enjoyment of their lands. The development of shale gas can include public nuisances such as increased noise, dust, traffic, odour, and visual impacts. The extent to which these affect people may depend on factors such as current environmental quality and the characteristics of the affected community including socio-economic status and community experience with the oil and gas industry.

The different stages of shale gas extraction require the transportation of equipment, chemicals, water, construction materials, and workers, usually by truck and often in large vehicles. A single hydraulically fractured well may need almost 2,000 one-way truck trips to deliver supplies, mostly water, and mostly during well completion. This is a concern not only for resultant noise and diesel emissions, but also because large truck traffic can damage rural roads that were not built to carry heavy loads, increase dust and congestion, and create an economic burden for local municipalities (Arthur et al., 2010). Truck traffic and road damage rank high among the concerns expressed by local residents in shale gas development areas in the United States (AER, 2011a). These areas have also seen increases in traffic accidents tied to the industry (Hill, 2013).

The sources of noise during shale gas extraction include drilling and hydraulic fracturing equipment, natural gas compressors, traffic, and construction. Drilling a shale gas well typically takes four to five weeks, 24 hours a day compared to about one to two weeks for conventional gas development (NYSDEC, 2011). As many as eight or more horizontal wells may be drilled sequentially from the same pad extending the total drilling time period to several months. Additionally, because hydraulic fracturing requires more pressure and water, more pumps and other noise-producing equipment are used (Arthur et al., 2010; NYSDEC, 2011). Noise can cause high blood pressure and other physiological effects, including sleep disturbance (WHO, 2009).

Shale gas development can also cause significant visual impacts on the local landscape (see Figure 7.2). Visual impact includes new landscape features: fencing, site buildings, land clearing, and well construction, for example (NYSDEC, 2011). The construction of well pads, roads, and their associated features, may also have a visual impact and are more long-term. How residents perceive visual impact depends on the local conditions, such as the value held for the landscape or the proximity to residential area, and also on temporal factors such as the time of day or year or the stage of shale gas extraction (NYSDEC, 2011). Furthermore, because drilling and completing a well is a 24-hour operation, lights have been identified as a significant source of visual impact on quality of life (Fraser Basin Council, 2012).

Figure 7.2. Shale Gas Development (Courtesy of www.marcellus-shale.us/).

Odours associated with various products used by the natural gas industry rank among the most common complaints made by local residents. They can be associated with health symptoms such as nausea, dizziness, headaches, and respiratory problems (Steinzor et al., 2013). Soil vibrations from fracturing may also change the colour, turbidity, or odour of well water.

7.1.6. Protection of Public Health

A series of recommendations related to shale gas development in that province. Three of these referred to the protection of public health related to changes in the local physical and social environment:

- The preparation of an equity focused health impact assessment (EFHIA).
- Protocols for monitoring the health status of people who live, work, attend school, or play in proximity to the industry are to be implemented.

- Linking health status information with environmental monitoring data and with data on socio-economic status.
- To include the assessment of short-term, cumulative, and long-term health impacts on the general public and any vulnerable populations.[15]

It is therefore recommended to include considerations related to vulnerable and disadvantaged populations that are at greater risk from environmental contaminants in planning and regulatory decisions. It also identified the importance of periodic reporting of environmental and health monitoring data to the public.

7.2. ETHICAL ISSUES

Some of the risks to human health, such as the slow migration of some contaminants in groundwater, the increased risk of cancer as a result of exposure to air emissions, or the intergenerational impacts of endocrine disruptors, have long latency periods and may affect future generations more than the current one. This argues for taking a long-term perspective when considering the health effects of shale gas development (Korfmacher et al., 2013). Some of these risks may also affect vulnerable populations disproportionately (e.g., individuals with pre-existing health conditions or without the means to avoid adverse impacts).

7.3. KNOWLEDGE GAPS

It can be concluded that the environmental and public health risks of shale gas development could not be quantified. This is in part because of

---

[15] The EFHIA, as opposed to a traditional health impact assessment (HIA), brings to light issues of equity when evaluating impacts to health as a result of project development.

uncertainty over the pace, scale, and location of development and the paucity of pre-development baseline studies.

The following gaps in knowledge of the effects of large-scale shale gas development on human health can be identified:

- The mixtures of chemicals associated with shale gas activities are generally unknown and untested, making it difficult to predict and assess risk from direct/indirect exposures.
- Concentrations of additives will change due to reactions with chemicals in shale-producing formations and dilution with brine. These reactions may produce new chemicals of potential health concern.
- The pathways of fracturing chemicals in the environment, including the routes through which individuals may be exposed, are unclear.
- Typical exposure duration times and concentration of different contaminants have not been fully established and specific health impacts are therefore difficult to predict or identify.
- Calculations of additive risk for specific compounds through different routes of exposure, or of cumulative risk from several compounds are not available.
- Public health surveillance, leading to epidemiological studies or rigorous health impact assessments of shale gas extraction activities have not been conducted.
- The lack of baseline monitoring has made it difficult to distinguish between ambient pollution and incremental pollution from shale gas activities.

Chapter 8

MONITORING AND RESEARCH

The preceding chapters conclude that the scientific basis for assessing the environmental impacts of shale gas development is weak, largely due to insufficient environmental monitoring. However, the problem extends far beyond mere data deficiencies. If a decision were to be made today to proceed with substantial monitoring programs, their effectiveness would be limited because a good understanding of how best to monitor for most of the potential impacts does not exist. The environmental monitoring needs specific to shale gas are, in important ways, different from those for other industrial activities. There is a need for research to determine how monitoring should best be done with respect to several of the potential impacts.

This chapter outlines broad principles for monitoring and describes approaches to monitoring impacts on health, gas emissions from the subsurface, seismic monitoring, monitoring surface water impacts, and groundwater monitoring. This chapter emphasizes monitoring of groundwater identified the potential contamination of groundwater resulting from shale gas development as being a significant threat to the environment. What is more, groundwater affects surface waters in many situations, and its contamination may also affect human health.

130   *Afsoon Moatari-Kazerouni*

This chapter also addresses monitoring objectives and the limitations in sampling domestic water wells to establish baseline conditions for the groundwater environment. Additionally, the status of mathematical methods for simulating and predicting subsurface impacts of shale gas development is examined because of their importance for understanding impacts. Also outlined is the relevance of aquifer vulnerability studies to a decision-making framework for groundwater protection.

The challenge of establishing a suitable framework for producing credible research results. It is essential that this framework involve government, industry, and academia and be balanced between issues of national importance and regional or provincial relevance.

8.1. MONITORING PRINCIPLES

The starting point for developing monitoring approaches is to define the principles that underpin such a program. Developing an effective monitoring plan according to these principles requires multidisciplinary research to provide a science-based framework as well as credibility.

Box 8.1. Principles that Underpin an Effective Monitoring Program

- *Holistic and comprehensive:* a systemic approach that incorporates multiple essential components of the system as well as the relationships among the components, integrates multi-scale spatial measurements and recognizes the temporal dimension, from past to future.
- *Scientifically rigorous:* a science-based approach that uses robust indicators, consistent methodology, and standardized reporting, including peer-review, that will result in independent, objective, complete, reliable, verifiable, and replicable data.
- *Adaptive and robust:* an approach that can be evaluated and revised as new knowledge, needs, and circumstances change and that ensures stable and sufficient funding.

Monitoring and Research 131

> - *Inclusive and collaborative:* an approach that engages concerned parties in the design and execution, including the prioritization of issues and setting of ecosystem goals.
> - *Transparent and accessible:* an approach that produces publicly available information (in forms ranging from raw data to analyses) in a timely manner that will enable concerned parties to conduct their own analysis and draw their own conclusions and that will make the basis for judgment and conclusions explicit.

Overall, Ewen et al., (2012) recommended the following monitoring categories:

- monitoring of leakage at the well and in pipelines;
- monitoring the hydraulic fracturing process including use of chemicals;
- monitoring of methane emissions to determine GHG footprint;
- monitoring of groundwater to allow major leakage to be immediately detected (ideally supported by an appropriate emergency plan that can be rapidly implemented);
- seismic monitoring, to help understand and prevent earthquake-based risks; and
- monitoring of well construction to determine structural defects manifest before a hydraulic fracturing-induced seismic event.

8.2. MONITORING OF HEALTH AND SOCIAL IMPACTS

Industrial developments, such as those associated with the oil and gas industry, have potential health and social implications. The type, frequency, and severity of any health and social impacts are highly dependent on the nature, frequency, magnitude, and complexity of development and on the geographic location and the physical, economic, and social environments in which the development takes place. People living in areas that experience rapid and large-scale development of oil and gas production using hydraulic fracturing are particularly at risk in terms of both social and health impacts (see Chapter 7). While there is little systematic research on either health or social impacts of hydraulic fracturing, examples highlighted in the relevant literature describe the importance of having a robust and comprehensive health-monitoring system in place before significant shale gas development occurs.

Health impact assessments (HIAs) are seldom required as part of the regulatory approval process by any provincial, national, or multinational jurisdiction. Mandatory and comprehensive requirements for submitting HIAs would have to be established provincially. Assessments should evaluate short-term, cumulative, and long-term health and social impacts, and consider mechanisms for enhancing health equity and the unique health and social needs of vulnerable populations.

Shale gas development is occurring largely in the traditional territories of who depend on the local environment for food and water and whose culture may be particularly affected. Specific monitoring of impacts on physical and mental health, social well-being, quality of life, and ecological systems on which they depend, is therefore essential. This includes not only impacts of shale gas development directly on their health, communities, and cultures, but also indirect and long-term impacts of intrusion into traditional territories and economic and social activities. Suitable HIA methodologies do exist. However, these methods are not being utilized.

Public health outcomes are influenced by a number of environmental, social, and economic determinants. These outcomes will be optimized when policies in each of these areas are complementary and can serve as the foundation for a Health Impact Assessment.

A HIA would estimate long-term cumulative health and social benefits and costs. In addition, the framework could include mechanisms to enhance health equity in development projects. Equity-focused health impact assessments (EFHIAs) are now being promoted internationally, (see Chapter 7 for more details). To ensure that all determinants of physical health and social well-being are assessed (see Chapter 7) with input and inclusion of all necessary governmental, institutional, provincial, municipal, and community agencies (see Figure 8.1).

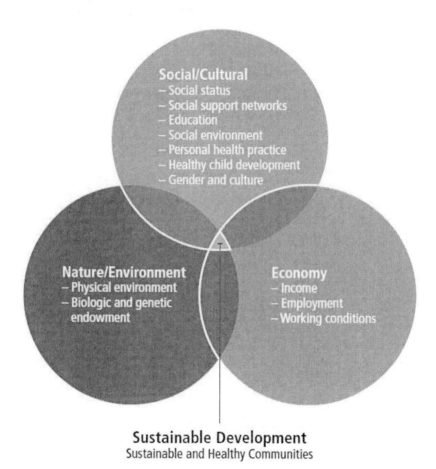

Figure 8.1. Health Impact Assessment (HIA) Framework (Reproduced with permission from the Public Health Agency Minister of Health, 2014).

8.3. Monitoring of Gas Emissions

Gas escaping into the atmosphere from wells with leaky seals has been identified as an important problem in shale gas development (as discussed in Chapters 4 and 5). Not only does this gas contribute to GHG emissions, but emissions at the surface indicate leakage below ground that could be going into freshwater aquifers.

Established methods measure gas escape at the wellhead, but these methods result in incomplete measurements because they do not include all the ways gas can potentially escape from the subsurface, which are:

- the well pad itself, made up of compacted fill material;
- the vadose (unsaturated) zone in the geological deposits beneath the fill; and
- below this, the groundwater zone.

The gas may migrate up the interior of the well inside the surface casing and therefore can be monitored in the wellhead at the top of the casing. Gas may also migrate outside the surface casing along the exterior of the cement seal around the conductor casing. Gas migration outside the conductor casing may follow pathways in permeable (mostly horizontal) geological layers to escape into the atmosphere at the periphery of or far beyond the well pad. Escape directly upwards through the well pad at points beyond the conductor casing is less likely because of low vertical permeability layers in the compacted fill within pad construction.

8.3.1. Testing Gas Leakage at the Wellhead

Natural gas emissions from the subsurface are typically measured as a surface casing vent flow (SCVF) at the wellhead. Complementing SCVF is the gas discharging outside the surface casing into the soil, referred to as *gas migration* or seepage (see Figure 8.2 and Figure 8.3).

If bubbling is observed, then determining the flow rate and the long-term pressure buildup is necessary.

Monitoring and Research 135

A gas leak to be serious if consider:

- hydrogen sulphide (H_2S) is present in the gas;
- the stabilized SCVF is at a rate higher than 300 cubic metres per day;
- the SCVF at stable shut-in pressure is greater than either 50 percent of the formation leak-off pressure at the surface casing shoe or 11 kiloPascals per metre times the surface casing setting depth;
- the SCVF includes oil;
- the SCVF includes substances dissolved in water that could contaminate soil or groundwater; and
- the SCVF occurs where usable or non-saline water is not protected with a cement sheath.

(B.C. Oil and Gas Commission, 2013b)

The EPA flux chamber is a static chamber with a real-time analyzer for oxygen, carbon dioxide, and methane measurements so that data are continuously monitored. It has become a standard tool for measuring gaseous emissions from land surfaces to the atmosphere or into buildings.

This chamber is the device most commonly used to measure gas escape across the ground surface at many types of contaminated sites, including petroleum pipeline leaks, service stations, industrial sites (where volatile halogenated chemicals exist in the ground), and most recently at locations of shale gas development.

Other types of devices and methods have also been used at contaminated site investigations as alternatives to the EPA flux chamber. For example, direct push profiling allows the collection of gas and water samples from one hole, simultaneously determining gas concentrations above and just below the water table (Amos & Blowes, 2008). Another similar method has previously been used in existing groundwater monitoring wells to obtain soil gas samples (Jewell & Wilson, 2011).

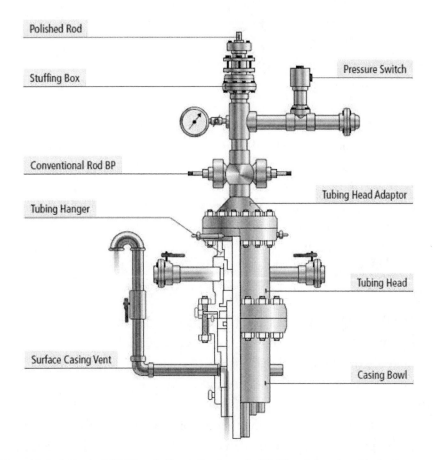

Figure 8.2. A Typical Wellhead. Cross-Section of a Wellhead showing the Surface Casing Vent (reproduced with permission from Theresa Watson. Watson & Bachu, 2009).

The apparatus used to carry out a bubble test on a wellhead. The numbers indicate:

1. wellhead,
2. surface-casing vent,
3. hose connecting surface-casing vent to water-filled container,
4. container with hose set beneath water surface to detect gas bubbles,
5. gas migration test hole, and
6. hand pump to direct accumulated gas to the lower-explosion-limit meter.

Methods such as these, developed in a hydrocarbon contamination context, could be applied to shale gas development particularly at and around the drilling pad. Using these and similar methods to assess background vadose and shallow water table gas conditions allows any changes induced by hydraulic fracturing and well failure to be detected.

Figure 8.3. Use of the Bubble Test to Detect Surface-Casing Vent Flow and Gas Migration. The apparatus used to carry out a bubble test on a wellhead. The numbers indicate: (1) wellhead, (2) surface-casing vent, (3) hose connecting surface-casing vent to water-filled container, (4) container with hose set beneath water surface to detect gas bubbles, (5) gas migration test hole, and (6) hand pump to direct accumulated gas to the lower-explosion-limit meter (courtesy of Theresa Watson. Watson & Bachu, 2009).

8.3.2. Testing Gas Leakage in the Vadose Zone

Conventional monitoring wells can be used to monitor gas migration in the vadose zone beneath and/or beyond the well pad. Options developed for other types of gas and groundwater monitoring can be used in shale gas monitoring. These methods produce profiles of gas concentration with depth at each monitoring location. The profiling of gas concentrations downward from near ground surface to the water table provides the best prospects for detecting gas migration. Geological layers of different permeability are common in the vadose zone and lateral gas migration is likely along more permeable layers, which may be thin with unpredictable distribution.

Gas migration in the permeable layers in the vadose zone is typically rapid. The most effective way to detect near-surface gas leaks beneath the

138 *Afsoon Moatari-Kazerouni*

well pad but outside of the surface casing of wells would therefore be profile monitoring in the vadose zone along the periphery of the pad. A layer of relatively high permeability only a few centimetres thick may be the primary pathway for gas to migrate from beneath the well pad to other areas. Gas profiling at petroleum spill sites is commonly conducted using bundles of tubes, each of a different length, sealed in a borehole with annular seals between each tube. Versions of these profiling devices are suitable for monitoring in the vadose zone and also in the groundwater zone (e.g., Einarson & Cherry, 2002). Research to determine the spatial scale at which profile monitoring will be most effective is still needed.

8.3.3. Other Air Monitoring

The sampling described above is only directed at quantifying leakage rates. Other types of gas sampling can be used to determine the specific health-related chemicals in emissions arriving at receptor points away from the well pad. Devices known as SUMMA canisters have been used to collect ambient air samples to assess air quality near pads in Colorado during well development, completion, and production (McKenzie et al., 2012) (see Chapter 7). However, this type of sampling is rarely done.

8.4. Seismic Monitoring

Regions considered for shale gas development would need to be studied for natural background seismicity, in a manner similar to that proposed by the B.C. Oil and Gas Commission (see Box 6.1). This background monitoring would form part of a local geological and seismic investigation to identify existing fault and seismicity patterns. This form of enhanced seismological network might involve 10 to 20 additional seismometers aimed at recording smaller magnitude events to delineate background levels. They could also form the basis of an ongoing local regional seismic network to detect changes in local seismicity if shale gas development proceeds. This seismic monitoring is distinct from that carried out by industry operators to map hydraulic fractures by monitoring and locating the induced microseismicity created by the fluid injection processes, as described in Section 6.2.

8.5. SURFACE WATER MONITORING

Shale gas development could potentially impact surface water in areas where sufficient disruption in the land causes changes in runoff of rainfall and snow melt, resulting in floods, erosion, and water quality problems.

Impacts specific to shale gas development are not well-defined, but many can be inferred from other regional developments. Effective monitoring programs are needed, yet there is little experience in doing this successfully. Sufficient baseline data for watershed indicators are uncommon. Good baseline data are required, and not all baseline data are publicly available. Linkages between surface water and groundwater are seldom sufficiently understood. A substantial surface monitoring program is required to understand the impacts.

8.6. GROUNDWATER MONITORING

Shale gas activities can be considered a major industrial operation because of the significant financial investments and the complexity of the activities incurred at each well pad. The scale of this investment and the related risks are inconsistent with the minimal environmental monitoring of air, soil, and water now conducted at each well pad.

Unlike hydraulic fracturing chemicals, pesticides must be tested to determine their propensity to cause groundwater contamination and must meet specific standards aimed at minimizing this risk before they can be approved for general use.

Some types of industrial, municipal, and agricultural activities can potentially contaminate groundwater and yet are not required to conduct site characterization and monitoring until an adverse event (e.g., spill or leakage of hazardous chemicals or well contamination) is suspected or occurs. This event then triggers a site investigation aimed at understanding the impacts and addressing mitigation. Landfills for municipal, industrial, or nuclear wastes, major chemical manufacturing facilities, sewage lagoons, and a few others do require site characterization and surface/groundwater monitoring systems for site licensing. However, what should be required for shale gas well pads is not clear. On the one hand, hazardous hydraulic fracturing chemicals and hazardous wastewaters, often radioactive, are stored at each well pad. Each well pad is therefore similar to landfills and chemical manufacturing facilities in terms of the risks posed to groundwater. On the other hand, the hazardous substances do not remain at each well pad for long, typically less than several months.

To date, no jurisdiction has imposed regulations requiring substantial groundwater characterization or monitoring at pads. This diminishes public confidence and limits possibilities for determining the cause of impacts reported for domestic wells. Whether characterization and monitoring should be done at all pads, just at selected representative pads, or just at well pads in areas designated as vulnerable to groundwater contamination, has not been addressed in the literature or regulations.

It considers appropriate to view the activities taking place at each shale gas well pad as a major — albeit temporary — industrial operation, because each well pad has an abundance of engines, fuels, hazardous chemicals, and hazardous wastes.

The challenges involved in shale gas development are clearly different for the surface and shallow subsurface environments and for the deeper Intermediate Zone. More is known about surface waters (i.e., streams, rivers, lakes, and wetlands) and the FGWZ because these parts of the hydrological cycle have been well studied in terms of other types of environmental impacts. This knowledge and many of the investigation methods and technologies can be applied to shale gas development. The differences posed by shale gas development mostly relate to the particular chemicals (i.e.,

Monitoring and Research 141

hydraulic fracturing chemicals and wastewaters). The literature on many or most of these chemicals in surface water and shallow groundwater is minimal. Therefore, the challenge is to adapt the existing framework of understanding for surface and shallow environments to accommodate the suite of chemicals most relevant to shale gas. There is an important role here for contributions from environmental chemists and geochemists collaborating with hydrologists, hydrogeologists, geologists, and geophysicists.

The Intermediate Zone is a much larger challenge. Very little is known about this zone concerning shale gas development. Questions mainly relate to issues around the occurrence and characteristics of natural fractures in the various types of rock strata and the related issues of migration of gas and saline water. The challenges with learning about the Intermediate Zone relate mainly to the greater difficulties imposed by higher temperature and fluid pressures, salinity, and the higher costs associated with drilling to the depths from which data are necessary. The technologies that could acquire data from this zone exist, but experience is limited.

8.6.1. Characterization and Conceptual Models

Modern professional practice analyzes environmental impacts at industrial sites in two stages: characterization and monitoring. Characterization provides the knowledge framework needed for the design of monitoring. Overall magnitude of studies needed at any particular CCS site exceeds that for any individual shale gas area, the general framework established for CCS is relevant. Characterization and monitoring are performed to:

- advance understanding;
- provide evidence for this understanding; and
- assess performance and mitigation actions

The need for more and better monitoring in shale gas development is often discussed in the literature. The essential prerequisite role of characterization is, however, usually not acknowledged. For some questions, conventional knowledge and practice are all that are needed to accomplish both characterization and monitoring, but for many of the unanswered questions about shale gas, further research is needed to inform these two stages.

Characterization refers to investigating the current nature and complexities of the groundwater system to understand migration pathways, identify receptors, and develop conceptual models that represent the entire system. Characterization can include data acquisition over time (temporal monitoring) to determine system behaviour. In subsurface studies, the characterization stage is commonly used to develop a conceptual model for flow and contaminant transport now and into the future. The conceptual model is often used as the framework for mathematical models to simulate the groundwater flow system and various scenarios for contaminant behaviour. Monitoring cannot be accomplished effectively and be scientifically defensible without first adequately characterizing the geology, hydrogeology, and hydrogeochemistry. A system cannot be monitored everywhere. Therefore, potentially affected groundwater systems need to be understood by developing conceptual models so that the monitoring devices can be positioned where impacts are most likely to occur.

In the quest to establish the occurrence and nature of groundwater contamination at industrial sites and provide the knowledge base for decisions about site restoration and remediation, emphasized the need to construct reliable conceptual models for subsurface conditions. This includes the geology, hydrology, and hydrochemical conditions and their variability in space and time in relation to understanding migration pathways and attenuation between sources and receptors. In normal professional practice, such characterization involves drilling to install down-hole devices that obtain groundwater samples and measure water levels (hydraulic head). The conceptual model embodies all that is known about the groundwater system to serve as the framework for decisions about additional characterization and monitoring (EPA, 1993; Kresic & Mikszewski, 2012).

CAPP (2012b) summarizes the desired outcome for shale gas investigations as "collaborating with government and other industry operators [...] to broadly understand regional groundwater quality and quantity through monitoring programs or studies that reflect good judgement and sound science." In the framework presented here, this endeavour is, in essence, the development of the conceptual model for the groundwater system.

Chapter 4 described the FGWZ contamination in two categories: that originating from surface-related activities on the well pad and related transportation accidents, and that originating from below due to leaky well seals and other deep pathways. The challenges for characterization and monitoring these two source zones differ. Although there is no literature specifically about the impacts on groundwater from surface sources at well pads, published literature on monitoring in the FGWZ is extensive and there are many similarities with the potential impacts of leaking containers, ponds, or landfills that have been extensively studied.

The monitoring approaches and understanding that have advanced markedly in the past three decades for other industries offer the framework for monitoring for surface sources in shale gas development.

8.6.1.1. Early Opportunities for Data Acquisition

In standard practice, geophysical borehole logging and microseismic investigations to assess the gas resources before intensive shale gas development do not include learning about the FGWZ or the upper part of the Intermediate Zone. In this practice, data acquisition starts when the drill hole penetrates a substantive depth below the FGWZ. However, some companies are beginning to assess shallower zones during this pre-gas production stage. For this, rock coring and geophysical logging of the holes are started at a shallower depth to include at least the bottom part of the FGWZ and the upper part of the Intermediate Zone.

Another opportunity to gather data is through microseismic holes that are typically drilled into the upper part of the Intermediate Zone but no deeper. These holes are normally only used for monitoring the hydraulic fracturing response. However, they could also be used for core examination and geophysical logging to study the shallower zones most relevant to fresh groundwater impacts.

System characterization can also take place while gas well drilling is in progress. The researchers used these fingerprints to match those sampled from domestic wells and, in some cases; the matches established the source depth zone of methane/ethane. This method is also applicable to shale gas development. This is a low-cost approach for the study of gas origins and, in addition to helping answer questions raised by domestic well complaints, the resulting information can be helpful where re-sealing of wells found to be leaking gas is required because the sealing can focus on the most likely leakage intervals. Repairing the leaky seal in a gas well can cost hundreds of thousands of dollars, and any information that helps to focus the repair work offers cost savings.

8.6.2. Monitoring Objectives

Many reviews on the environmental impacts of shale gas development recommend greater attention to monitoring of groundwater. However, none indicates how such monitoring should be done or even provides strategies for doing so. First, a framework to determine how much monitoring is required needs to be established. The overarching principles for monitoring and relevance to research are indicated in Box 8.1. Here, the objectives examines for groundwater monitoring.

Along with compliance monitoring and research monitoring, monitoring can also be directed at any or all of the following objectives to establish baseline conditions and identify effects in particular parts of the system:

- performance monitoring;
- sentry monitoring; and
- receptor monitoring.

Performance monitoring aims to determine the degree to which the industrial activity is performing as intended. For example, in the context of shale gas development, such monitoring would concern not only the degree to which gas leakage is occurring along the cement-seal annulus and outside, but also along the annulus and beyond the well into the groundwater system. If the well is performing as intended, the leakage will be non-detectable or below a specified limit. Other examples include monitoring at or close to a pad to observe contamination emanating from beneath the pad, and monitoring above the hydraulic fracturing zone to determine whether the hydraulic fracturing has caused gas or saline fluids to migrate upward due to out-of-zone fracture propagation. Performance monitoring is also relevant to demonstrating no leakage from the tanks containing the hydraulic fracturing chemicals and the tanks or ponds holding the flowback water. An objective of shale gas development is to minimize all of these leakages, and this monitoring is therefore directed at determining whether the operation is performing as intended.

Performance monitoring incorporates practices and procedures to achieve the stated goals and objectives of industrial operations. It includes worker safety, occupational health, public health, and environmental impacts. When monitoring is configured within a regulatory framework, it is known as *compliance* monitoring and may include components of performance, sentry, or receptor monitoring.

Sentry monitoring is done between the source and the receptor along most-likely potential migration pathways for early detection before arrival to the receptor of interest. There may be several receptors of interest and several migration pathways. A sentry-monitoring network can improve its efficiency through planning to ensure it detects multiple pathways and receptors of concern as frequency of sampling will depend on migration pathway lengths, the particular contaminant being examined, mobility, and attenuation mechanisms. Sentry monitoring is about detection before arrival to the sensitive or valued receptor — and the issue being addressed is evidence of mobility and possible threat to the identified receptors, and whether corrective measures need to be taken to reduce the impacts or remove the threat(s). Since there are multiple pathways, receptors, and constituents of concern, an integrated systems approach based on the best available science is likely to be most cost-effective, with variable timelines and triggers for additional tiers of sampling.

Receptor monitoring, on the other hand, is direct monitoring of the resource itself (e.g., the aquifer, the municipal or private well) to determine if changes and degradation above an acceptable standard have occurred (health, environmental, or aesthetical). Receptor monitoring is done to ensure the impacts have not occurred to a level of concern. However, health, environmental, or nuisance standards for air, groundwater, or surface water do not yet exist for many of the constituents used or mobilized during shale gas development. Receptors should include the resources themselves (air, aquifers, and surface waters) and may include but not be limited to existing water supply wells (public and private) and springs. To assess impacts to air, groundwater, and surface or well waters, baseline sampling is needed to be able to assess impacts or changes in water quality. Baseline sampling is not a sufficient metric, and the frequency and timeline for impact analysis may require many decades after development has commenced and post abandonment.

To accomplish each of the three types of monitoring (performance, sentry, and receptor), baseline monitoring must focus on each category to provide the basis for identifying changes over time attributable to shale gas activities and, therefore, involves monitoring devices positioned in different parts of the groundwater system.

8.6.3. Methods of Monitoring Groundwater

Technologies and strategies for groundwater monitoring have advanced markedly during the past decade. Many options are available to choose from depending on the problem at hand. Monitoring groundwater requires devices installed in drill holes. Traditionally, groundwater monitoring was done using monitoring wells established as conventional practice in the early 1980s. However, conventional wells have been replaced by multilevel monitoring systems (MLS) to monitor many types of groundwater problems.[16] MLS are a more effective way to use each drill hole as it allows much more information to be collected. Drilling costs are generally the factor limiting the detail provided by groundwater monitoring networks and, therefore, monitoring wells generally result in sparse data because each hole provides only one monitoring point. In contrast, MLSs provide data profiles with depth so that critical zones for contaminant migration are more likely identified. These profiles are initially used for characterization but can also be used for later long-term monitoring.

There are many examples where MLS have been used for characterization and monitoring of groundwater impacts for various types of industrial activities (e.g., Meyer et al., 2008; Chapman et al., 2013), including upstream oil and gas development. However, no published references found to MLS being used to detect the impacts of shale gas development.

CAPP was one of the first organizations to recommend the use of monitoring wells for monitoring well pads (CAPP, 2012b). That said, recognizing the advances made in groundwater monitoring in the past three decades, the term *monitoring well* should be taken here as a generic term to mean whatever type of device is most appropriate to serve the monitoring goals of each borehole. Installing a few conventional monitoring wells will likely not be effective for well pad monitoring because the most critical depth zones for contaminant migration will probably be missed due to geological complexities.

A vision for shallow monitoring at a well pad (e.g., to depths of about 20 to 40 metres below ground surface, depending on local conditions) could include a few MLS for profile monitoring positioned along the periphery of the pad. Some sampling points in the MLS would be positioned in the vadose zone and some in the groundwater zone so that the presence of gas migration from beneath the well pad and groundwater quality can both be determined at some or all monitoring locations. Gas migration in the vadose zone is caused primarily by diffusion and thus typically occurs in all directions from

---

[16] See Patton & Smith (1988) for a general conceptual rationale for the use of MLS rather than conventional monitoring wells for groundwater impact monitoring. Different types are described in Einarson & Cherry (2002) and Einarson (2006).

the gas leakage points. Consequently, profile monitoring would need to be conducted on all sides of the pad. The shallowest groundwater flow will be directed by the slope of the water table. However, any seasonal variability in this slope may not be evident until the monitoring devices are in place. Therefore, monitoring on all sides of the well pad would ensure that all contaminants would be detected regardless of the direction of the flow. Prior site characterization to determine a dominant groundwater flow direction is marginally useful since the dominant direction may vary seasonally or from year to year.

Monitoring at well pads is further complicated by not knowing the chemical composition of the flowback water, which may vary from pad to pad and particularly from area to area due to disparate shale characteristics. Even if the complete chemical composition of the hydraulic fracturing substances is known, the chemical composition of the flowback water cannot be fully characterized as it comprises a mixture of injection fluids and shale water with significant levels of salts, metals, and, in some cases, NORM. Furthermore, little is known about the persistence, fate, or toxicity of these chemicals (including the biocides in the injection fluids) when subjected to the high temperatures and pressures manifest in shale formations. Laboratory research into the breakdown and transformation of hydraulic fracturing chemicals, including hydrolysis, biodegradation, and mineral-induced transformations, is in its early stages (e.g., Kahrilas et al., 2013). Ultimately, laboratory results will be needed to corroborate field data.

Given the large degree of uncertainty about the chemical composition of both the hydraulic fracturing chemicals stored at each well pad before use and the flowback water, monitoring the groundwater for many of the chemicals is not feasible. Therefore, groundwater monitoring will likely require initial reliance on indicator parameters (tracers) that would signal the presence of hydraulic fracturing chemicals or flowback water. When changes in indicator parameters are detected, more comprehensive analyses could be initiated. Suitable indicator parameters have not yet been identified in the literature.

Monitoring and Research

8.6.3.1. Mathematical Simulations and Models for Groundwater Systems

A near universal component of understanding and predicting long-term future behaviour of fluid movement in subsurface systems is the mathematical representation or modelling of fluid flow and contaminant reactive transport. Mathematical simulators are useful tools to gain insight into the behaviour of complex hydrogeological systems by developing models of specific systems, such as an aquifer or a drainage basin. The simulators are the linked collection of numerical algorithms that represent the dominant processes being reproduced. The model output is an attempt to reproduce some site-specific observed or predicted environmental behaviour with that simulator. In the context of shale gas development, a model of gas flow up the annulus between the cement seal of a shale gas production well and the rock surface of the borehole would need to consider the physical pathways and the driving forces of gas migration through brine, subject to pathway heterogeneities and temperature and pressure variability. A model that mimics gas migrating into a shallow aquifer as a result of such an event must allow for both gas and groundwater to flow in a permeable medium and account for gas buoyancy and dissolution into groundwater. Multiphase fluid flow simulators for gas and water are regularly used by hydrogeologists and petroleum engineers. However, the specific goals can differ and cause each discipline to make different assumptions. Modelling the biogeochemical reactions between methane gas and groundwater requires a geochemical reaction simulator. Such simulators are also now in use to model various reactive transport situations. No simulator yet exists that can account for (i) annular gas flow, (ii) aquifer invasion, and (iii) geochemical reactions. Leaky wells and their effect on groundwater can only be approximated piece-wise rather than modelled as an integrated, fully coupled system.

Robust models that make reliable predictions require extensive empirical data that can be compared to model predictions, adequate timeframes, and the ability to compare model output and actual conditions. In hydrogeological science, numerical models have been applied extensively over the past few decades to predict groundwater flow and contaminant transport in relatively shallow aquifer systems. Although most of these models have been tested and compared with observed data, many do not perform very well (Oreskes & Belitz, 2001).

In the case of shale gas development, hydrogeologists have neither the history of scientific data collection that would allow them to construct robust models with well-characterized input parameters, nor a broad empirical basis against which to compare model predictions. Significant advances have been made over the past two decades in modelling complex systems, including multiphase systems (mixtures of gases and aqueous and non-aqueous phase liquids), heterogeneous and fractured media, and biogeochemical reactive transport.

Modelling subsurface flow in shale gas environments is not yet practical, primarily due to a lack of basic scientific data on the nature of fracture networks and a relatively poor understanding of fluid flow in low permeability rocks, especially under dynamic rock stresses and transient fluid conditions.

State-of-the-art multiphase flow and transport simulators suitable for modelling hydrogeological processes in the context of the local environment of shale gas development include TOUGH2 and TOUGHREACT (Pruess et al., 1999; Pruess, 2005; Xu et al., 2012), DuMU^X (Flemisch et al., 2011), and COMPFLOW (Unger et al., 1995). These types of codes provide the mathematical means to simulate gas and liquid-phase systems under various conditions, and have been used effectively in production environments. Predicting resource extraction outcomes versus the migration of low- or trace-level chemical quantities presents different challenges because of the large differences in spatial and temporal scales. Due to the lack of necessary field characterization data, these models will not reliably predict long-range or long-term impacts of shale gas development on regional groundwater resources. Various reservoir engineering codes, including GEM (CMG Ltd.) and ECLIPSE (Schlumberger Ltd.), also have multiphase processes. However, they were designed primarily for optimizing hydrocarbon production, not for predicting environmental impacts. If the impacts of shale gas development on groundwater quality are to be understood through simulation, simulators capable of coupling gas and fluid-phase migration with dissolved-phase geochemical reactions and geomechanical processes will have to be developed and applied in conjunction with adequate characterization and monitoring data on appropriate scales to define these models on a site-or area-specific basis. This has not been done to date, likely because it is an immense challenge with respect to both data acquisition and computation.

Conceptual hydrogeological models and physical processes related to geologic carbon dioxide sequestration have some similarities to the context of shale gas development. Much can be learned from these applications, and many of the conceptual models and numerical approaches could be adapted to simulate groundwater impacts of shale gas development with an emphasis on sedimentary rock with formation sequences, including layers of very low and much higher permeability strata with fractures. Birkholzer et al., (2011) and Zhou et al., (2010), for example, use the TOUGH2/ECO2N simulator (Pruess, 2005) to simulate carbon dioxide injection into deep saline aquifers and evaluate effects on shallow groundwater systems. To date, this has only included hypothetical scenario testing for evaluating heterogeneities of key processes. Lemieux (2011) includes a review of numerical model applications to carbon dioxide storage and notes a similar lack of data for model calibration.

Numerical modelling of hydrogeological systems related to shale gas extraction has thus been limited and simplified. Ewen et al., (2012) used numerical models with literature-derived input data to examine leakage of hydraulic fracturing fluids upwards into the FGWZ in an area of Germany targeted for shale gas development. Simulations were used to guide judgment on safe vertical distances between the shale gas zone and the bottom of the FGWZ. Gassiat et al., (2013) simulate brine transport along permeable faults using literature-based ranges of physical and hydraulic properties but necessarily make a series of assumptions and simplifications. Myers (2012), for example, applied the United States Geological Survey model MODFLOW to simulate several potential fluid pathways from a shale gas deposit to ground surface, in the context of the Marcellus Shale in New York. Simulated scenarios with fault pathways suggested that transport of fluids from a shale gas reservoir to ground surface could be on the order of few decades or less. Saiers and Barth (2012) identified several critical shortcomings of Myers' approach, however, including neglecting the gas phase, neglecting density effects of formation brines, and neglecting the effect of higher temperatures with depth. The limited scale and boundary condition constraints were also noted. Vidic et al., (2013) also indicate that Myers' model includes numerous simplifications that compromise its conclusions. Cohen et al., (2013) indicate that the results of this modelling exercise were pre-determined by Myers' assumptions and boundary conditions, and that the model was meaningless without field data to support its conclusions. These critiques are consistent with the overall conclusion that, to date, a scientifically accepted model or suite of models has not yet been developed to predict, for realistic field conditions, the impacts of shale gas development on regional groundwater resources where contamination can be caused by gas leakage or other constituents migrating from below into the FGWZ.

8.6.4. Monitoring Contamination from Below the Fresh Groundwater Zone

Developing monitoring approaches to detect contamination originating from below the FGWZ is much more difficult than doing so for surface sources. The literature indicates that the pathways for migration from below at shale gas sites have not been investigated for any type of contaminant. The main contaminant of concern is methane, potentially originating in gas-rich strata in the Intermediate Zone and the deeper shale gas zone.

The scenarios for the migration of gas and saline water from below up into the FGWZ include many variations, the simplest being where the gas migrates along a leaky well annulus and then moves mostly horizontally once entering permeable strata in the FGWZ. For gas-permeable pathways, more complex scenarios involve potential horizontal migration in deeper permeable (i.e., fractured) strata in the Intermediate Zone connected to the well annulus, and then upward into the FGWZ. From a monitoring perspective, experience cannot suggest the most likely form of a gas migration pattern or plume. If the gas plumes are very narrow because they only follow a particular fracture or thin permeable bed, they will likely not be detected by the monitoring network because financial constraints limit the spacing between monitoring points to some practical limit. However, if the gas is laterally dispersed (e.g., fan shaped, spreading transverse to the main direction of migration), then the monitoring target will be much broader and easier to detect. Based on typical plume shapes of other industrial contaminants in fractured sedimentary rock (e.g., Parker et al., 2012), a gas plume emanating upward from the fractured rock in the Intermediate Zone is likely to be quite dispersed and therefore have a shape and size that is favourable to detection by monitoring.

Jackson et al., (2013a) reviewed a sample of domestic wells, which suggested that methane attributable to hydraulic fracturing pads within one kilometre was present in the well water (as discussed in Chapter 4). The researchers then used additional geochemical fingerprinting techniques to narrow uncertainties about the origin of the methane (Vengosh et al., 2013). They concluded that:

> the elevated methane in drinking water wells near the shale gas wells had a thermogenic composition (e.g., heavier ^{13}C-CH_4) than wells located 1 [kilometre] away from shale gas sites with an apparent mixed thermogenic-biogenic composition. New emerging noble gas data reinforce the carbon isotopes and hydrocarbon ratios data and indicate that the high levels of methane exceeding the hazard level of 10 [milligram per litre] are indeed related to stray gas contamination directly related to shale gas operation. The most probable mechanism for stray gas contamination is leaking through inadequate cement on casing or through the well annulus from intermediate formations.

Without strong transverse dispersion during gas transport by groundwater, the likelihood of gas from a well pad showing up in many domestic wells is small because the gas plumes would be narrow and much less likely to encounter domestic wells.

154 *Afsoon Moatari-Kazerouni*

Given that well leakage is the main threat of contamination from below, there is a need for monitoring devices that can be implanted along the surface casing and perhaps even along the conductor casing to detect gas leakage directly or measure indirect indicators such as fluid pressure or perhaps temperature. The petroleum and geotechnical industries have the technology and expertise to measure such parameters but have not yet focused these widely enough on the problem of well leakage. The most desirable technology would be one that indicates the depth of the leakage so that the well can be repaired where the threat warrants remedial action.

8.6.5. Domestic Wells

Whether or not shale gas development impacts groundwater is a matter of debate. Nearly all of the evidence for or against has come from domestic wells, with no efforts directed at verification using other more reliable groundwater monitoring devices.

There are two reasons for sampling domestic wells: to determine the quality of the drinking water used by the well owner as a baseline to assess future shale gas impacts, and to gain understanding of the groundwater system as part of system characterization. The people who rely on domestic wells share a concern that shale gas activities will affect their wells. Therefore, sampling should be done to address this concern, but this sampling alone should not be accepted as more than simply a component of the baseline groundwater monitoring framework.

A domestic well typically has a steel casing sealed in a borehole with cement that extends from the surface down to the top of the well intake (i.e., screened) interval, generally 10 to 100 metres below ground depending on the local geology. When pumped, the water enters the well from the various geological layers or zones that constitute the aquifer. The water pumped from the well is therefore a mixture of shallower and deeper groundwater that may have substantially different chemical compositions. In some areas, the deeper groundwater may be more likely to contain dissolved natural methane. The nature of this mixture for any specific well can vary depending

Monitoring and Research

on the season, the timing of the last rainfall, the rate at which the well is pumped before the water sample is taken, the position of the pump intake in the well, and other factors. This makes interpreting the sampling results from domestic wells a challenge.

The typical design of domestic wells in many areas makes them vulnerable to contamination from surface sources. Shallow wells in unconfined aquifers (i.e., those not buried beneath geological deposits with low permeability) and wells in bedrock are particularly vulnerable because the surface casings of these wells are typically not deep enough to protect against contamination. Furthermore, cement, which deteriorates over time (Lackey et al., 2009), is commonly used to seal the casing of domestic wells. Robust sampling of domestic wells commonly shows the effect of leakage along the casing from surface runoff (e.g., bacteria are often found in domestic wells) or shallow groundwater short circuiting to deeper zones. Not only is this a potential threat to the health of the well owner, such leakage can complicate identifying shale gas impacts. Leakage along the casing of domestic wells can cause sampling of domestic wells to vary over time, making it more difficult to identify the factors that influence water quality.

Private well owners rarely have their well water tested for more than bacterial contamination. As a result, they typically know little about the quality of the water aside from aesthetic indicators (e.g., smell, taste, turbidity). Other than the study by Gorody et al., (2012), no other published comprehensive assessments of domestic well sampling for implications about baseline conditions. Sampling a domestic well once or twice and then drawing the conclusion that the quality is normal or acceptable is dangerous in the public health context because infrequent sampling may give results unrepresentative of the typical drinking water quality. Overall, domestic well sampling, when done rigorously, can be of value from the perspective of drinking water. However, this sampling needs to be combined with other essential sampling elements as part of baseline monitoring programs aimed at understanding the fresh groundwater resource.

The most common issue concerning shale gas impacts on groundwater is natural gas, particularly methane, found in domestic wells. Debate continues over the sources of the methane — is it natural or attributable to shale gas drilling and the hydraulic fracturing of these wells? Natural methane is common in the FGWZ and therefore methane presence on its own is not proof of a relation to shale gas development or to any other oil and gas development. The essential issue then is how to distinguish existing methane from any methane contributed by shale gas activities. The simplest conceptualization of the problem is that existing methane originates at shallow depth and is biogenic, usually geologically young methane. In contrast, the methane generally attributable to oil and gas industry drilling is much older thermogenic gas, coming from the Intermediate or Deep Zone due to leaky well seals or other short-circuit pathways. Isotopic analyses have been used successfully to distinguish these source types (e.g., Tilley & Muehlenbachs, 2011).

8.6.6. Aquifer Vulnerability

Aquifer vulnerability is a concept used for groundwater resource management and protection against many types or causes of groundwater contamination. It typically implies threats from surface sources (Focazio et al., 2002; Brouyère et al., 2011) though this concept is also relevant in the context of risks to groundwater posed by shale gas development. Vulnerability refers to the susceptibility of an aquifer to contamination. It depends on many factors including the types of contaminants and the way in which they are most likely to enter the aquifers. Methodologies to determine the development of aquifer vulnerability in the context of shale gas development are complicated by three factors specific to this industry: the lack of information on the assimilation capacity of aquifers for fugitive gas emissions and for many of the chemical constituents in hydraulic fracturing fluids and in wastewaters; the potential for contamination of the aquifers via pathways from below; and the anticipated high gas-well densities and associated risk of cumulative long-term impacts on aquifers. In situations where aquifer vulnerability has been used as a component of groundwater protection, the contaminant sources have been at or near land surface and therefore were relevant to assessing the potential impact on groundwater of surface releases at shale gas well pads.

If government agencies were to anticipate impacts of shale gas development on aquifers and to apply management practices on a region or area-specific basis, some form of aquifer vulnerability classification specific to risks from this development would be required. Ideally, an aquifer vulnerability assessment methodology would provide a framework for deciding about designation of areas as too risky for shale gas development and for selecting safe separation distances between shale gas well pads and existing domestic wells and other potential receptors of impacts. Applying such a methodology would also contribute to identifying priorities for monitoring. Vulnerability also depends strongly on the type of aquifer, such as whether it is unconfined or confined and whether it is granular (e.g., sand or gravel) or fractured rock. It also depends on the background flow regime and the nature of aquitards present separating contaminant sources from receptors including those in the Intermediate Zone. A confined aquifer has at least one overlying geological layer of low permeability; surface contaminants are less likely to enter the aquifer quickly or without much prior assimilation. Unconfined aquifers are generally more vulnerable to surface contamination.

Frind et al., (2006) use these concepts to define *intrinsic* and *specific* aquifer vulnerabilities, with *intrinsic* referring to vulnerability due to aquifer properties, and *specific* also including factors such as whether organic chemical contaminants are entering the aquifer as oily liquids or as dissolved contaminants (solutes). Equal attention must also be given to aquitards and their properties that allow either contaminant migration or attenuation. Aquitard integrity is an important part of assessing aquifer and well vulnerability, but aquitards deeper than those assessed in the past are now becoming more relevant in the context of shale gas development.

A related concept is well vulnerability, which considers the complete pathway of a contaminant to a water supply well (Frind et al., 2006). The concept of well vulnerability is applicable to domestic wells in areas of shale gas development. In many jurisdictions, municipal wells have wellhead protection areas around them; that is, the land uses are restricted to activities least threatening to groundwater quality. For example, within the normal framework of wellhead protection schemes, a shale gas well pad would not be permitted within the capture zone of a water supply well. Set-back restrictions for the deep horizontal well sections are less clear. Methods for determining aquifer and well vulnerability to surface contaminant sources with pathways to shallow unconfined or confined aquifers are well known and include index methods (e.g., DRASTIC; Aller et al., 1987) and physically based numerical models (e.g., Frind et al., 2006), respectively. Applications to fractured rock aquifers are more difficult to determine, but new approaches have been proposed (e.g., Pochon et al., 2008).

Programs for mapping aquifer vulnerability (at least at regional scales) and delineating protection areas for municipal wells are substantially advanced in most jurisdictions where groundwater is an important source of drinking water. Vulnerability maps can be developed for all major aquifer systems at the (municipalized) watershed scale, including within the primary shale gas target areas. To define aquifer vulnerability and wellhead protection areas accurately, substantial additional effort is nevertheless required to characterize aquifer systems and develop hydrogeological site conceptual models. The substantial progress made in aquifer vulnerability assessment methods in the past several decades may be useful as a framework suitable for shale gas development, which introduces new complexities due to threats from below. No methodology specific to the needs of shale gas development currently exists. Field characterization for defining vulnerability from below, for example, will be much more challenging given the increased depths and potentially more complicated contaminant pathways including faults and fractures. Existing aquifer vulnerability maps for potential surface sources are also typically prepared at regional or sub-watershed scales and are therefore not often sufficiently detailed to be useful at the local scale of a shale gas well pad.

Ewen et al., (2012) concluded that shale gas development "can entail considerable environmental risk, particularly when it comes to water resource conservation which we strongly feel absolutely must take precedence over energy production." This report recommended that hydraulic fracturing should not be permitted in three types of areas deemed

Monitoring and Research 159

too vulnerable: geologically unstable areas, pressurized artesian and confined deep fresh groundwater and permeable faults, and areas already designated for high priority groundwater protection.

8.6.7. Separation Distance Issues

In shale gas development, there are important concerns about horizontal distances between the gas well and other wells, such as domestic water wells, springs, municipal wells, or old oil and gas wells. This is especially important in regions where there is a high rural population density, such as in the target shale gas region, which averages a few dwellings per square kilometre, each with their own domestic water well. Some jurisdictions have regulations that specify minimum distances between domestic wells and gas wells. For example, 200 metres as the minimum distance can be defined between a shale gas well and the nearest water supply well (See Table 9.1 in Chapter 9). However, setback distances are arbitrary in that they are not based on scientific analysis related to likelihood of impacts. Several factors are involved, such as the vulnerability and attenuation capacity expected of the local area in question, and the characteristics of the water supply and other wells. What is low risk in one area, therefore, may be high risk in another depending on the hydrogeological situation. In the event of a claim of damage made by the landowner, the operator is held responsible. The presumptive liability distance is now specified as 2,500 feet (760 metres), compared with the previous distance of 1,000 feet (300 metres) (Pennsylvania Office of the Governor, 2012 as referenced in Coussens & Martinez, 2014).

Although this assignment of presumptive responsibility establishes the legal framework and provides increased assurance to landowners with respect to compensation, it does not do much to protect the potable groundwater resources or to provide better understanding of when and how impacts may occur. No guidance found in the literature indicating what a scientific framework for establishing minimum separation distances could be.

The vertical distance between the base of the FGWZ and the top of the zone subjected to hydraulic fracturing is also a concern. This vertical aspect is important in shale gas development because it concerns minimal vertical distances necessary to minimize contamination of freshwater aquifers from below. At what depth below groundwater should hydraulic fracturing be considered too risky? King (2012) indicates that 2,000 feet (600 metres), or within 1,000 feet (300 metres) of fresh water, is shallow enough to warrant special consideration to make sure that conditions are suitable. Ewen et al., (2012) commissioned mathematical modelling studies (albeit without site-specific field data) to examine this question and concluded that the distance between ground surface and target gas reservoirs should exceed 1,000 metres. In addition, the distance between the bottom of the deepest fresh groundwater and such reservoirs should exceed 600 metres (Ewen et al., 2012). However, these depths are intended to avoid contamination of fresh groundwater resources in demonstration projects subjected to intensive monitoring and research so that minimum depth questions can be answered on an area-specific basis. The studies did not identify types of monitoring needed to determine safe separation distances.

8.7. RESEARCH ACTIVITIES AND OPPORTUNITIES

The environmental impacts of shale gas development from the United States, United Kingdom, and Australia all identify strong needs for research (SEAB, 2011a, 2011b; The Royal Society and Royal Academy of Engineering, 2012; ACOLA, 2013). Other reviews have also identified this need, such as Ewen et al., (2012) which recommended a research strategy focused on "proceeding cautiously, one step at a time so as to allow for careful testing and ensure that hydrofracking is not pursued in haste." This report also recommended that shale gas development initially proceed in a limited manner, using demonstration projects to "enable scientists to study in greater depth the impact of widespread use of hydrofracking, in light of the surface and sub-surface conditions." Further, they recommended that demonstration projects be intensively monitored, focusing on the geomechanics of fracture propagation, well integrity, and contaminant occurrence, transport, and fate (Ewen et al., 2012).

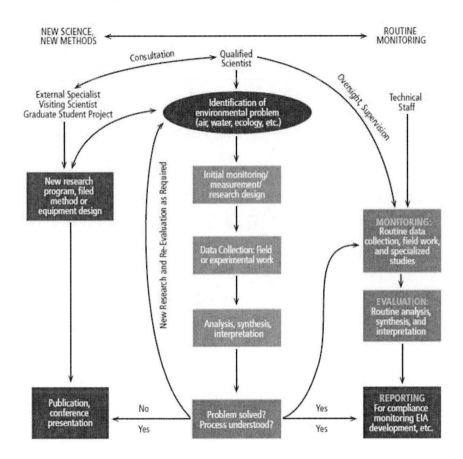

Figure 8.4. The Role of Science in Informing Environmental Monitoring. Illustration of good practice for science-based environmental monitoring (adopted from AEMP, 2011).

To date, to the best of the knowledge no country with the exception of the U.S. EPA study has initiated substantial subsurface, field-focused research at an actual hydraulic fracturing well pad to examine impacts and related processes beyond those looking at secondary research questions and retrospective case studies. Without such research, important questions will remain unanswered. Rigorous monitoring is needed as part of the research process and also over the longer term to assess cumulative impacts and provide evidence to the public that environmental effects are as claimed, stated that:

> governments must deal with the inherent conflict of being the resource owner, regulator and revenue taker. For the [...] monitoring system to have the requisite legitimacy and scientific credibility, the system must operate at arm's length for all affected parties, including governments, regulators, and those being regulated. [...] To be credible, the information that arises from a monitoring program must be regarded by stakeholders as scientifically sound and free of bias or perceived bias. Science must drive the design, execution and evaluation of monitoring programs [...]. Transparency contributes to credibility and will be a hallmark of monitoring program and the data information they produced.

Moreover, they indicated that "an essential element of credibility is that the output of the monitoring system (data and reports) can withstand review and scrutiny by independent scientists" (AEMP, 2011). On the topic of research credibility it states that:

> there is substantial evidence that trust in political institutions and processes have diminished over the last two to three decades. Government is no longer seen as an independent arbiter [...]. There are also low levels of trust towards multinational corporations and big business [...]. This means that information coming from these sources is likely viewed [skeptically] by many in society [...]. [Research independence] can be addressed, at least in part, by having guidelines for the use of industry funding (e.g., rights to publish clauses, peer review requirements) and by being transparent about funding arrangements and process whereby projects are selected and signed off.

For shale gas research to be credibly viewed by the public and the scientific community at large. Transparency, so important for public confidence, is one of the basic requirements of university-based research. If the research community is to fill knowledge and technology gaps, government, academia, and industry will all need to participate.

As a result, and because the most likely scenarios for potentially substantial impacts largely differ from those for which monitoring experience exists, the first requirement is to develop a science-based framework for characterization and monitoring. This will require field-focused research at shale gas areas and it indicates that key components of world-class science include:

> scientists trained in methods of observation and inquiry. In the modern world, the system of graduate education that leads to the doctoral degree (Ph.D.) has been designed to generate such individuals. Increasingly, post-doctoral experience under high level academic supervision is required before an individual is considered suitable to take a scientific leadership role [...]. It is an essential part of the process of science for scientists to consult freely and widely with peers, colleagues and specialists outside their immediate circle. This ensures the maintenance of high standards and also ensures that the scientist is in touch with the latest developments in his/her field, which improves the efficiency, productivity, accuracy or relevance of the results being generated. Peer-reviewed publications are part of this process, and result in the science being fully transparent.

The nature of the research considered needed to overcome the lack of knowledge about well leakage and the associated groundwater impacts. There are two parts to the problem. The first concerns the well seals, leakage pathways, and rates of leakage into aquifers and escape at surface. The second concerns the impacts that the gas can have on fresh groundwater resources. The first is mostly about physical processes and the second mostly about geochemical and microbial processes that cause *in situ* natural gas attenuation. By drawing on experience in related fields, it is evident that a comprehensive research program would combine laboratory experiments, mathematical modelling, and short- and long-term field investigations and experiments. For well-integrity issues to be understood and mitigation measures to be considered reliable, there will need to be convincing field demonstrations with rigorous monitoring and publications in peer-reviewed journals. The ExxonMobil funded report from Germany recommended intensive research, initially at few test sites in a small area in Germany (Ewen et al., 2012) which is, in essence, a field laboratory approach. In this approach, characterization and monitoring methods can be developed and tested while also improving understanding of potential pathways and mechanisms of fluid migration. Such a field laboratory approach has been used successfully in many other areas of environmental research. Field laboratories provide the observational data essential to advance conceptual and mathematical models for understanding and predicting impacts on larger spatial and temporal scales. The results of the field laboratory stage are subsequently used to inform the nature and intensity of site characterization and monitoring appropriate for application at hydraulic fracturing sites following the initial test stages.

8.7.1. Research Activities and Opportunities

Modestly funding several university-based teams to summarize the state of knowledge in five areas related to hydraulic fracturing and water:

- "Water use and demand management.
- "Wastewater handling, treatment, and disposal.
- "Groundwater and subsurface impact issues.
- "Landscape impacts of development/operations on surface water/watersheds.
- "Watershed governance and management approaches for resource development, including Aboriginal issues."

The administrative frameworks that include combinations of government and industry funds, and are suitable if research on the

Monitoring and Research 165

environmental impacts of shale gas development is mobilized. These existing research programs make it readily possible for industry and government funds to be used jointly by academics collaborating in groups of substantial size with transparency and publishing in appropriate scientific journals. This type of research also plays a major role in educating and training the younger generation of researchers who can focus on important long-term questions.

8.7.2. Provincial Research

The summaries below indicate how each of four of the main shale gas provinces in is responding to actual or prospective shale gas development. Each has recognized science and technology gaps and has developed research or is in the process of developing mechanisms or plans for research to fill some of the gaps concerning water characterization or monitoring. The nature and likely magnitudes of the impacts will differ somewhat between provinces and even between areas within a province. As such, some research topics are best viewed as national concerns because they are generic, whereas others are more area- or site-specific and of provincial importance only. Well integrity is obviously an issue of national and international importance, but implementing characterization and monitoring for each of the shale gas areas is, or will be, a provincial activity. However, the magnitude of the sustained funding needed for rigorous well-integrity research is likely much beyond the resources that any individual province could commit. Effective approaches and technologies to conduct such characterization and monitoring need to be established in each province. However, some of the components used to determine what types of characterization and monitoring would be most effective, are common to all provinces; this therefore warrants joint attention. There is a challenge to create opportunities for research that uses substantial industry funds but nevertheless produces results that are highly credible in both the scientific community and the public view.

8.8. CONCLUSION

There are many issues concerning the environmental impacts of shale gas development for which there are no answers due to the current lack of adequate characterization, monitoring, and study. Research is just beginning to receive some attention in the United States, and research programs are in the earliest stages. An important issue not yet addressed concerns what questions should receive research priority and how should proceed to organize shale gas research.

To better understand the risks to surface water and groundwater resources, a significant commitment will be needed to develop effective baseline monitoring and apply effective operational monitoring. It is particularly challenging to implement a monitoring program for the cumulative effects of shale gas development that is sensitive to the watershed-scale. The cumulative effects are most significant at this geographic scale. In the face of development with incomplete knowledge, an adaptive monitoring and management philosophy emphasizing transparency would identify unanticipated impacts as soon as possible.

The need for post-production cumulative effects monitoring should also not be underestimated. Even when plans/procedures/monitoring programs are in compliance, unforeseen cumulative consequences of development may only be detected by an effective post-operational monitoring program and adaptive management.

What little monitoring has been done in areas of active shale gas development has used domestic wells. Although such sampling, if done comprehensively, can indicate baseline conditions for these particular wells, this is not an effective means for determining the baseline conditions or understanding the groundwater system. The challenges faced by government and industry concerning monitoring are large because it will be necessary to develop and test methods and technologies for monitoring and system characterization to determine what is effective and efficient before rigorous longer-term monitoring can be done. The challenge is compounded by the need for performance assessment, sentry, and receptor monitoring, which include both domestic wells and monitoring needs encompassing not only the freshwater resources being used by people now, but also potentially usable resources for the future.

There is much agreement in the literature that shale gas development will only gain broad public acceptance if the monitoring is done in a credible and fully transparent manner. However, there are no established examples of what such a monitoring program would look like. Another challenge is to ensure that research results are openly discussed and debated, while not delaying the use of robust results to improve methods and regulations. There is a need for rapid transfer of new understanding to practice so that benefits are realized quickly and efficiently.

For the research to be most credible and to take full advantage of the established expertise that exists, substantial funding from both industry and government will be required. However, the research results must remain credible. For credibility, the research should not be directed by industry, but industry advice and internal expertise will be essential to meet the scientific and technical challenge. There is a need for research that is generic, with results relevant to all shale gas provinces, as well as region- or area-specific research, with results of most relevance to individual provinces or specific areas within provinces. Given that shale gas environmental research is in the early stages, there is considerable opportunity to become a major player in this research domain.

To address the most important questions about the environmental impacts of shale gas development, it will be necessary to conduct collaborative research involving many disciplines at representative field locations before, during, and after hydraulic fracturing. Community study and monitoring will be needed to determine health and social impacts. In addition to the NSERC programs referenced above, it would be appropriate for social science-related research to access the recently created SSHRC programs called Partnership Development Grants and Partnership Grants. Similarly, the CIHR has an institute for Health Systems and Policy Research and another on Population Health that may have relevant programs.

For the non-health and social aspects, laboratory and field experimentation will be needed. For some questions, such as the deterioration of cement seals over time, experiments and environmental media measurements are required along with the development of treatment and monitoring technologies, involving in many cases long-term data acquisition. Most research projects are planned for the relatively short term — five to ten years or less. To address some of the most important questions about shale gas, longer-term research will be needed over the anticipated decades-long development periods and over sufficient time scales following well closure.

The purpose of the research is to inform government and industry about impacts so that, in response to the findings, improvements in monitoring, mitigation, and management can be implemented. For the research to have most value, processes must be in place to make results available transparently and in a timely manner. Given the urgency of developing science-based procedures and regulations to minimize the potential for long-term cumulative impacts, the transfer of information and technology to industry and appropriate government bodies is an important challenge.

Chapter 9

MANAGEMENT AND MITIGATION

Large-scale shale gas development poses a number of risks to the environment and human health, as the preceding chapters elucidate. The existence of knowledge gaps related to these risks underlines the importance of acquiring additional information through monitoring and research. This would allow the risks and their impacts to be analyzed, prevented where possible, and appropriately mitigated. In this review, the term *mitigation* refers to any engineering actions or alterations, regulations, policies, management plans and procedures, or changes in social organization and institutional culture aimed at reducing actual or potential environmental impacts of shale gas development.

On the issue of mitigation options, it is possible to comment on "the state of knowledge of associated mitigation options" and more specifically:

- What technical practices exist to mitigate impacts?
- What are the gaps in science and technology relevant to possible mitigation measures?
- What are international good practices to mitigate impacts?
- What science underpins current policy or regulatory practices internationally?

170 Afsoon Moatari-Kazerouni

Addressing these questions posed some challenges arising from:

- uncertainty over the location, pace, and scale of future shale gas development
- significant regional differences in geological, environmental, and socio-economic characteristics (e.g., population density);
- the continuing evolution in both the technologies used to develop shale gas (e.g., number of wells per pad, composition of fracturing fluids) and in the regulatory and policy frameworks to manage it;
- the paucity of baseline information and scientific understanding about key environmental factors (e.g., groundwater, fluid flow in low permeability rocks); and
- differences in public perception and acceptance of shale gas development across the country.

As discussed in earlier chapters, the impacts of shale gas development are difficult to quantify with respect to groundwater and surface water, air quality, land, human health, and community well-being because little is known about the pathways by which contaminants can move from the shale gas activities to receptors. Because of such limited understanding about these pathways and impacts, particularly long-term cumulative impacts, it is not possible to address the mitigation questions directed quantitatively. Nevertheless, the technologies employed in shale gas development are sufficiently advanced, as are many of the concepts about how environmental impacts may occur, to allow to identify general measures that would lessen the potential for some of these impacts. There are also methods or systems for risk and safety management which have been proven effective in other industries, and which believes are well suited for use in shale gas development.

Consequently, it is chosen to focus on selected science and technology gaps related to mitigation and the main attributes of a possible risk-management framework for shale gas development for two reasons:

Management and Mitigation   171

- to emphasize the overall architecture of such a framework rather than individual mitigation measures, the requirements for which can vary for historical, institutional, or regional reasons and must respond to technological changes and actual development proposals; and
- to recognize the cumulative effects of large-scale development.

The mitigation of adverse effects is not only a question of deploying appropriate technical practices but also requires regulators and operators to implement appropriate policies governing large-scale development.

Good practices are still evolving. Nevertheless, in this chapter a number of mitigation measures considered that could be expected to diminish potential environmental impacts of shale gas development.

The purpose of this chapter, consistent with the charge to lay out information relevant to how likely environmental impacts may be mitigated in areas where development proceeds. It recognizes that shale gas development is already underway in two provinces, but that arguments have been made in other jurisdictions about the benefits of developing slowly. It is not inherently obvious, for example, that extracting gas from a shale bed is better for society if it begins now and ends in 20 years rather than beginning in 3 years and ending in 23 (Goldstein et al., 2013), especially in cases where land use planning and the regulatory framework are incomplete. Gas-drilling technology is improving rapidly in ways that both decrease the likelihood of significant adverse incidents, such as well casing failures, and increase ultimate recovery. In other words, the pace of development is a key determinant of the success of a risk-management strategy.

The risk-management framework presented below is relevant to multiple development scenarios and that its application would be beneficial in a variety of contexts. While it is believed that effective measures exist to mitigate several of the impacts of shale gas development, there will still be impacts even with mitigation practices in place. However sophisticated or well-intentioned, government and industry managers cannot guarantee that all environmental risks will be alleviated or all impacts avoided if development proceeds.

9.1. FRAMEWORK FOR RISK MANAGEMENT

Based on the experience of other industrial sectors (e.g., the chemical industry), it is believed that the mitigation of the environmental impacts from the exploration, extraction, and development of shale gas resources rests on a comprehensive approach that focuses on five distinct, mutually reinforcing elements:

a. *The technologies to develop and produce shale gas.* Materials, equipment, and products must be adequately designed, installed in compliance with specifications, and reliably maintained.

b. *The management systems to control the risks to the environment and public health.* The comprehensive and rigorous management of materials, equipment, and processes associated with the development and operation of shale gas sites will ensure public safety and reduce environmental risks.

c. *An effective regulatory system.* Rules to govern the development of shale gas must be based on sound science, and compliance with these rules must be monitored and enforced.

d. *Regional planning.* To protect the environment, drilling and development plans must reflect local and regional environmental conditions, including existing land uses and environmental risks. Some areas may not be suitable for development whereas others may require specific management measures.

e. *The engagement of local citizens and stakeholders.* Public engagement is necessary not only to inform local residents of development but also to identify what aspects of quality of life and well-being residents value most, in order to develop a process that wins their trust and protects their values.

In this framework, each of these elements would be supported by environmental monitoring programs that include activities such as research, characterization, and modelling. The purpose of these programs is to detect changes associated with development, check the accuracy of predictions,

Management and Mitigation 173

assess risks, design mitigation strategies, and evaluate performance. The monitoring issues addressed in the preceding chapter.

9.2. TECHNOLOGIES FOR SHALE GAS DEVELOPMENT

The extraction and development of shale gas resources involve numerous activities and technologies. Many of these are well established in the oil and gas industry, as are many of the practices that protect environmental quality. Rather than review technical practices that mitigate environmental impacts in each of the activities involved in shale gas development, focuses below on one area that is essential to protect groundwater resources and minimize GHG emissions.

9.2.1. Well Construction

9.2.2.1. Findings on Well Integrity
Although it has identified well integrity as important, this is not a new concern with regard to the environmental impacts of shale gas development. Prior to this assessment, reports examined the environmental impacts of shale gas development for their respective governments: the United States (Secretary of Energy Advisory Board), the United Kingdom, and Australia (SEAB, 2011; The Royal Society and Royal Academy of Engineering, 2012; ACOLA, 2013). These reports provide recommendations and are broadly consistent in that they identify many of the same issues, although with substantial differences in emphasis. The advantage of reviewing these reports and their recommendations, and having access to substantial new scientific literature that has appeared since these reports were produced. The major issues identified in these reports concern groundwater and surface water impacts; ecology, land, carbon footprint, and climate change; health and community well-being; operational mistakes/accidents; and public relations and communications. Well integrity relates to many of these in one way or another.

The earliest of these national reports, from the United States, states that "inspections are needed to confirm that operators have taken prompt action to repair defective cementing jobs" (SEAB, 2011a). Similar sentiments are evident in the U.K. report, which states, "ensuring well integrity must remain the highest priority to prevent contamination" (The Royal Society and Royal Academy of Engineering, 2012). The U.K. 1996 *Offshore Installation and Wells (Design and Construction) Regulations* require that the design and construction of onshore and offshore wells be examined by an independent and competent person (not necessarily a third party) who can review the results of well integrity tests and raise health and safety concerns; however, they cannot prohibit activities (U.K. Parliament, 1996). In their report, the U.K. Royal Society and Royal Academy of Engineering (2012) recommended that the independence of this function be strengthened.

The Australian report probes more deeply into the well integrity issue with emphasis on the longer-term view, stating, "cement and steel do not have very long-term integrity in geological materials" (ACOLA, 2013). The report also indicates that there is a lack of adequate data and analysis to define what constitutes a failed well and states that:

> well abandonment is not just a regulatory issue but is also an issue that requires more research and development in areas such as the very long-term behaviour of cements and extended monitoring under hostile subsurface conditions […]. The very long-term integrity of a cemented and abandoned well (beyond 50 years) is a topic where more information will be essential.

All three of these reviews advocate well integrity monitoring over the life of the well.

The issue of well integrity was also identified by the ExxonMobil-funded in Germany, whose report states "the industry's eight decades of experience with the long-term stability of cement shows that gas well cementing does not remain leak-proof indefinitely" and "that abandoned and sealed wells need to be monitored so as to detect any gas or contaminant emissions early enough to take necessary countermeasures" (Ewen et al.,

Management and Mitigation

2012). In Chapter 3, the causes of cement deterioration were outlined and indicated that cement deterioration can be very slow. This raises the possibility of needing to monitor wells *in perpetuity* because, even after leaky older wells are repaired, deterioration of the cement repair itself may occur.

Recognition that wells may leak several decades or longer presents a challenge for all governments responsible for regulating shale gas development. The challenge involves balancing the desires of our current society for the economic benefits of this natural resource with the ethical imperative to avoid passing on the responsibility for well maintenance and impact monitoring to future generations.

Of the four reports discussed above, only the German report provides specific recommendations based on the long-term uncertainties, with the main one being to "take it slowly and tread carefully" and conduct long-term basic research (Ewen et al., 2012). It recommends that shale gas development should proceed with "a small number of hydro fracking projects being approved on a case by case basis and that they be scrupulously monitored by rigorous application of the scientific method." The initial few projects would serve as the field laboratory to address many of the uncertainties that have been identified, with the understanding that the results of this initial stage will determine the future of shale gas development (Ewen et al., 2012).

The U.K. Royal Society and Royal Academy of Engineering (2012) noted a European Sustainable Operating Practices Initiative for Unconventional Resources will "establish a 'field laboratory' to test and demonstrate best practices independently funded by entities not actively involved in oil and gas extraction." In its filing requirements for hydraulic fracturing, the goal of these requirements is to ensure that the operator demonstrates that:

- two or more independent and tested physical well seals (barriers) are in place throughout all phases of well operations;
- well seals ensure well integrity during the entire well life cycle, and under all load conditions, hydraulic fracturing and completion;

176 *Afsoon Moatari-Kazerouni*

- repairs are made or other action taken without delay if the well control is lost or if safety environmental protection or conservation of resources are threatened;
- the safety of the workers and population is maintained and that hydraulic fracturing will not cause waste or pollution; and
- all equipment is tested to the maximum pressure to which it is likely to be subjected.

9.2.2.2. Provincial Regulations

Another example of risk mitigation is stipulated separation distances between gas wells and domestic wells.

As discussed in Chapter 8, there is little scientific basis for these setback regulations. Less distance would increase risk, but the degree of risk reduction remains ill-defined in the absence of data. A clear need for the continued study of well integrity, and for flexibility in regulatory requirements that suit different geological conditions, is apparent.

Given the tens of thousands of shale gas wells that may be drilled in the next 50 to 100 years, there is also a need for credible research into these important questions, undertaken by research that is independent of influence from sectors with vested interests. The principal users of such information would be industry, but governments and their regulatory agencies would also benefit from the improved scientific and technical base for public policy aimed at improved safety.

9.3. RIGOROUS RISK AND SAFETY MANAGEMENT

Process safety management (including emergency response plans) is the management system that is generally used to ensure public safety and environmental protection (*viz.*, for example, OSHA 1910.119 and EPA Risk Management Program). The application of state-of-the-art technologies and practices can greatly reduce the environmental risks posed by shale gas development, but these risks do not arise only from the misapplication of technology. Inadequate management systems also contribute to

Management and Mitigation 177

environmental degradation; it is therefore important to ensure their performance as well.

There is evidence that some companies are upgrading their safety practices in light of the challenges posed by shale gas development. For example, the close spacing of completion operations on a single pad to reduce the overall environmental footprint of a development (and to save costs) creates unique safety challenges:

- the drilling of a well and the fracturing of another one on the same pad can occur both simultaneously and in close physical proximity;
- high-pressure stimulations could communicate in the subsurface with the drilling rig; and
- on a 16-well pad, there are 16 potential sources of pressure from the wells themselves in addition to the pumping equipment, each posing a safety risk

These factors have led to significant changes in the way some companies operate in the field, including the development of purpose-designed safety protocols to minimize these risks. The development of these protocols must take into account the fact that a single well may involve different contractors building the pad, drilling the well, cementing it, fracturing it, delivering supplies, and hauling wastes. To a greater or lesser extent, all are involved in ensuring the safety of the operation and all must subscribe to the agreed safety procedures.

As with shale gas development, other industries (e.g., oil sands extraction, chemical manufacturing, mining) use hazardous technologies, products, or processes; manage industrial wastes; and build permanent and temporary infrastructure. These industries have also developed systems to manage risks associated with these operations. Experience in these industries (*viz.*, various voluntary codes such as the chemical industry's Responsible Care or the mining industry's Towards Sustainable Mining) demonstrates that rigorous risk management needs to be based on the definition, quantification, and documentation of functional performance

178 *Afsoon Moatari-Kazerouni*

requirements and the reduction of risks to that which is as low as reasonably practical for the complete life of the site (Moffet et al., 2004).

More specifically, a rigorous risk and safety management system typically includes the following elements, although the nomenclature and presentation may vary across industries:[17]

- The creation and nurturing of an environmental protection and safety culture across the organization.
- A hazard identification and risk assessment process that characterizes the probability of an event happening and the severity of the potential consequences.
- A risk-management system that covers activities and responsibilities, such as:
  - management of change procedures to identify and control hazards associated with changes to facilities/standard operating procedures/staff to prevent the introduction of new hazards and keep the information up-to-date;
  - standard operating procedures, including a list of steps to follow in executing a given task, describing the way to execute each task and their sequence;
  - safe work practices for repetitive tasks;
  - contractor management to ensure that contractors are informed of the risks associated with a facility and the safety and environmental rules they need to follow;
  - training to provide facility employees and contractors with the information and skills they need to accomplish their tasks and protect the environment;
  - procedures to ensure that critical equipment and structures are designed, constructed, verified, inspected, monitored, and

---

[17] Such a performance system needs to be verified. The publication *Recommended Practice: Risk Management of Shale Gas Developments and Operations* provides a reference for third-party verification of management systems (DNV, 2013). The IEA also calls on operators to recognize the case for independent evaluation and verification of environmental performance (IEA, 2012b).

Management and Mitigation

179

    maintained as recommended by the manufacturers, applicable industry standards and as a function of their use;

- review of safety and equipment readiness before using new equipment or after extended shutdown;
- a formal process for incident reporting and investigation, including determining how and why each accident happened and the recommendations to prevent reoccurrence;
- an emergency response plan to ensure that the site staff is always aware of the risks and knows what to do if things go wrong — such a plan should be integrated with the local community and public authorities emergency response plans; and
- the identification of laws, regulations, and standards applicable to the facilities to be managed; regular audits to ensure compliance and the implementation in a continuous improvement process.

- A performance management system, including performance indicators to track process safety and environmental protection performance. Indicators can track process safety and environmental protection performance and allow the government, industry, and the public to "determine whether play development and project plans and any other related processes are effective in meeting regulatory outcomes, determine the effectiveness of regulatory tools and processes, identify where opportunities exist for improvement, proactively identify new risks or thresholds being approached or exceeded" (AER, 2012e).

A successful risk-management process will systematically adapt its practices by learning from outcomes. Such continuous improvement is embedded in the management practices of some industries (e.g., chemical manufacturing) and requires explicit decision-making processes to be effective.

9.3.1. Standards and Voluntary Codes

A wide variety of organizations have developed standards, codes, and guidance to embed risk safety management into the management systems of shale gas operators (see Table 9.1).

9.3.2. Standard-Setting Organizations

The American Petroleum Institute has developed many standards for the North American oil and gas industry, including some specifically related to hydraulic fracturing and shale gas development. Some international examples include the Det Norske Veritas (DNV) recommendations. In a related vein, a standard for the geological storage of carbon dioxide (such storage raises some of the same issues as shale gas, *viz.*, the possible uncontrolled migration of fluids in the sub-surface) (CSA Group, 2012).

9.3.3. Voluntary Codes

The oil and gas industry is also reviewing its practices to adapt them to shale gas development. An industry's environmental performance is guided not only by government regulations but also by social expectations, corporate culture, and financial considerations, among others. These factors are often embodied in voluntary codes, defined by industry as "codes of practice and other arrangements that influence, shape, control or set benchmarks for behaviour in the marketplace." Existing codes address a wide variety of environmental protection, consumer, community, and other issues and, while voluntary, can often have significant legal implications. These codes can be referenced in legislation, elaborate on regulatory requirements, or be used by the courts as a standard for the appropriate level of care in an industry (Webb, 2004).

Management and Mitigation 181

Voluntary codes relevant to the shale gas industry exist. While many address health and safety issues and some address quality management, others are designed to improve environmental performance.

CAPP, whose members control 90 percent of natural gas and crude oil production, has published a set of guiding principles and operating practices related to hydraulic fracturing activities that covers seven areas:

- fracturing fluid additive disclosure (CAPP, 2012h);
- fracturing fluid additive risk assessment and management (CAPP, 2012g);
- baseline groundwater testing (CAPP, 2012b);
- wellbore construction and quality assurance (CAPP, 2012f);
- water sourcing, measurement, and reuse (CAPP, 2012e);
- fluid transport, handling, storage, and disposal (CAPP, 2012d); and anomalous induced seismicity — assessment, monitoring, mitigation, and response (CAPP, 2012c).

Each of these guiding principles, developed by consensus among industry operators, includes a description of operational requirements, performance measures, and reporting expectations. Some also include flow charts and checklists. CAPP will report annually on the implementation of this voluntary code by its members, starting in its 2013 progress report.

The group also collaborates on road construction and maintenance, plans drilling sites around existing infrastructure, and uses roads as pipeline corridors, thus reducing environmental and development costs.

Table 9.1. Examples of Standards and Voluntary Codes Relevant to the Shale Gas Industry

Organization	Code
American Petroleum Institute	•• API HF1 Hydraulic fracturing operations — well construction and integrity guidelines
Det Norske Veritas (DNV)	•• Risk management of shale gas developments and operations
International Organization for Standardization	•• ISO 9001 Quality management systems •• ISO 14001 Environmental management systems •• ISO 31010 Risk management – Risk assessment techniques
Occupational Health and Safety Advisory Services	•• OHSAS 18001 Occupational health and safety systems •• OHSAS 1910.119 Process safety management of highly hazardous chemicals
Chemical Industry Association	•• Responsible Care commitments
Organisation for Economic Co-operation and Development	•• Corporate governance for process safety

9.3.4. Company Codes

Some companies have developed their own programs or operating principles to protect the environment. Encana's Responsible Products program aims to ensure that the products they use in their hydraulic fracturing operations are safe, effective, and as environmentally responsible as possible. Its Courtesy Matters program addresses nuisance issues such as garbage, traffic, noise, dust, and lease activities that affect local communities (Encana, 2013).

9.3.5. Third-Party Guidance

Social sector organizations have also developed recommended goals and practices for managing the risks associated with shale gas development (see Table 9.2).

9.4. EFFECTIVE REGULATORY SYSTEM

Effective government oversight of hazardous operations helps protect human health and the environment. A control system to ensure compliance with applicable laws and regulations usually includes a suite of measures such as rules, standards, prohibitions, mandatory reporting, inspections, and penalties. Systems can be prescriptive (e.g., by stating what technology is to be used) or performance-based (e.g., by specifying a desired outcome). In both cases, the standard of performance needs to be based on sound science (e.g., the level of emissions acceptable from a human health or environmental protection point of view).

Setting and enforcing such a standard is often difficult: the available science may be inconclusive, conditions may change making existing standards less effective, and resources may be insufficient. Maule *et al.* (2012) document the challenges faced by some U.S. states with extensive

Box 9.1. National Energy Board's Requirements

The *Filing Requirements for Onshore Drilling Operations Involving Hydraulic Fracturing* shows how a risk management system can be applied to hydraulic fracturing. It requires the preparation of five distinct plans to manage risks. Summarizes below the objectives of these plans, each of which includes detailed filing requirements.

(i) Safety Plan
"The Safety Plan should provide enough details to demonstrate the procedures, practices, resources, sequence of key safety-related activities, and monitoring measures necessary to ensure the safety of the proposed work or activity."

(ii) Risk Assessment and Risk Management Plan

- "The [Risk Assessment and Risk Management Plan] describes the applicant's risk assessment and risk management processes with enough detail to demonstrate that the applicant has:
- effective processes in place to identify threats and hazards to safety and the environment, to identify and select effective environmental measures and to evaluate and manage the associated risks; and
- taken, or will take, all reasonable precautions to ensure that safety and environmental protection risks have been addressed for the proposed work or activity, including taking into account the interaction of all components, including structures, facilities, activities, equipment, operating procedures, and personnel."

(iii) Environmental Protection Plan (EEP)

"The EPP should provide enough detail to demonstrate:

- an understanding of how the work or activity will interact with the environment;

- that the EPP has the procedures, practices, resources, and monitoring necessary to manage hazards and protect the environment from the impacts of the proposed work or activity, including potential impacts to groundwater; and that the predicted environmental hazards and risks, including the preventive and mitigative measures in the [Environmental Assessment], are incorporated."

(iv) Waste Management Plan

"'Waste material' is any garbage, refuse, sewage, waste well fluids, or any other useless material that is generated during drilling, completions, hydraulic fracturing, formation flow testing, well or production operations, including drill cuttings, used or surplus drilling and completion fluids, hydraulic fracture fluids, produced fluids including formation fluids and flowback fluids. Applicants are expected to take all reasonable measures to minimize the volumes of waste materials generated by their operations, and to minimize the quantity of substances of potential environmental concern contained within these waste materials. No substance should be discharged to the environment unless it has determined that the discharge is acceptable [...]. Applications must include a complete and adequate plan to manage discharged waste material."

(v) Spill Contingency Plan

"A spill contingency plan should provide emergency response procedures to mitigate environmental and safety impacts from unplanned or accidental discharges to the environment. Pollution, which includes spills, also refers to situations where discharges from authorized operations or activities exceed the authorized discharge limits [...]. Applications Contingency Plans for spill response will provide enough detail to demonstrate that effective systems, processes, procedures, and capabilities will be in place to:

- minimize the impacts to the natural environment from unauthorized or accidental discharges; and
- protect workers and the public."

shale gas development in implementing an effective compliance assurance system related to disclosure of fracturing chemicals (e.g., inadequate agency budgets, understaffing, and deficient records). These challenges can be exacerbated when the pace of development accelerates, straining the capacity of regulatory authorities to monitor environmental changes or enforce regulations (U.S. GAO, 2012a; Vaughn, 2012). In its *Golden Rules for a Golden Age of Gas*, the IEA (2012b) notes the importance of ensuring that regulatory bodies receive adequate resources, including sufficient permitting and compliance staff.

9.5. REGIONAL PLANNING

The third element in the framework for managing the effects of shale gas development on the environment and human health is regional planning, the need for which flows from two distinct characteristics:

- Shale plays vary greatly in terms of geology, the local environment, and social conditions; mitigation and monitoring measures need, therefore, to reflect this variety and be regionally specific.
- Relative to conventional gas, the greater scale of development and concentration of infrastructure required to produce shale gas imply increased land impacts and land use conflicts; the only effective way to manage such cumulative effects is at the regional, not local, scale.

The need to take a regional approach to managing the cumulative impacts of shale gas development (e.g., AER, 2012e; IEA, 2012b). In its proposed new rules, the AER recognizes that effective planning can reduce the amount of infrastructure needed and make it more efficient (AER, 2012e). Such planning can also improve the management of the environmental effects of development. The AER is, therefore, looking to encourage operators in a shale gas play to create a common development plan for regulatory approval, collaborating on addressing issues such as water management, surface infrastructure, and public engagement. To facilitate placing multiple horizontal wells on a single well pad location, the AER is also proposing a new pad approval process (i.e., multiple well approvals) to supersede individual well approvals (AER, 2012e). Increasing the number of wells per pad reduces the number of well pads required to drain the reservoir as well as the number of roads and utility corridors, thereby reducing surface impacts. While each multi-well pad is bigger than single-well pads, the overall land footprint will be smaller because fewer pads will be needed (ACOLA, 2013).

The following land use plans can be developed to determine resource access and management:

- Land and Resource Management Plans (LRMPs) that "define the location of protected areas, special management zones, enhanced development and general management zones with supporting objectives and strategies."
- Sustainable Resource Management Plans "that translate the strategic or higher level objectives from LRMPs into more specific resource management direction required for day-to-day operational decisions."

Oil and gas development is allowed in special management zones subject to specific conditions.

This moratorium recognizes that there may be areas where shale gas development is not suitable, either because there is insufficient environmental information, or because the land use conflicts resulting from development cannot be resolved at an acceptable cost (e.g., destroying the habitat of an endangered species; destroying culturally significant land). Given the environmental risks that shale gas development raises, it believes that there will be specific areas where development would raise unacceptable costs.

Box 9.2. How to Manage Cumulative Effects

The AER has identified the following practices to help manage the cumulative effects associated with unconventional oil and gas development.

- Collaborate early on at the play level to anticipate infrastructure needs, examine low-impact options, and implement plans that balance environmental, social, and economic needs.
- Collaborate on use and siting of new and existing infrastructure to minimize proliferation and heavy truck traffic.
- Maximize use of existing infrastructure such as roads, well sites, and pipeline corridors.
- Collaborate with other industries on roads and use of already disturbed sites.
- Collaborate on operational matters to reduce risk (e.g., traffic accidents, road-use plans, offset well communication).
- Support use of pads as the new standard and address any regulatory issues that may constrain their use.
- Adopt best practices and use the best available technology to mitigate effects of noise, lighting, and dust.
- Address end-of-life infrastructure liability effectively without any cost to the public.

(AER, 2012e)

It is recognizes that implementing a regional approach to development to reduce cumulative effects will require a significant investment in human and financial resources. The many examples of the environmental costs of not addressing the cumulative effects of other forms of development up front (e.g., alienation of agricultural land through urban sprawl, over-fishing, excessive nutrient loads in lakes, disappearing wildlife habitat) underline the importance of taking a regional approach when considering the large-scale exploitation of shale gas.

In a related vein, the IEA (2012b) is encouraging regulators and operators to take advantage of economies of scale that become available for some activities as the scale of development increases. A larger scale creates opportunities for some mitigation options that would not be economical at a low rate of development. Some of the mitigation practices that become available with economies of scale are:

- drilling multiple wells from single pads;
- using pipelines rather than trucks to carry water, thereby reducing truck traffic;
- centralizing water management and treatment facilities;
- reducing venting and flaring because wells can be connected to a pipeline after completion; and
- replacing mobile diesel engines with fixed electric or CNG engines, thereby reducing air emissions

In addition, economies of scale enable more detailed geophysical mapping and environmental monitoring to be undertaken because costs can also be spread out over more wells. Investing more money in reservoir characterization allows operators to focus development on *sweet spots* that contain more resources, rather than drill at regular intervals. The IEA estimates that a better understanding of geological structures and hydrocarbon flows in part of the Barnett Shale might have avoided the drilling of the least profitable 20 percent of the wells that were drilled (IEA, 2012b). Better geological information could therefore result in appreciable financial savings and fewer environmental impacts. However, historical experiences with the mining and petroleum industries raise doubts as to whether this is likely to occur. For financial and other reasons, most operators proceed as soon as they acquire minimally adequate geological information (Oreskes, 2011).

9.6. PUBLIC ENGAGEMENT

Over the past few decades, engaging the public in decision-making concerning major development initiatives has become normal — if not required —— practice. Recent experience across different industries demonstrates (e.g., pipelines, wind energy, waste management) the value to proponents of engaging local citizens, particularly when they are deploying technologies with which they may not be familiar. Where the rights of Aboriginal Peoples may be affected, governments also have a constitutional obligation to consult them.

From a proponent's perspective, a goal of public engagement is to gain local support for and acceptance of a project. As Shindler and colleagues (2002, 2003) point out, earning this acceptance is often a function of how those affected by a project perceive the legitimacy of the decision-making process: if they do not trust the proponent or the government to protect their interests, they are unlikely to provide their support, regardless of the project's merits. As the experience of shale gas development, public acceptance is also situation-specific: practices that are acceptable in one situation may not be in another. A public engagement strategy needs to reflect these differences and be oriented to local context, capacity, and concerns.

Individuals make judgments about the acceptability of a project based on a series of factors, including issue salience, personal values, previous experiences, knowledge of the situation, the quality of information, beliefs about the fairness of the process, trust in decision-makers, and risk perceptions. As conditions or information change, they may re-assess their judgment and proponents need to respond with appropriate management actions. Winning and maintaining social acceptability therefore requires an ongoing effort (Shindler et al., 2002).

Public engagement ideally involves a dialogue between the promoter and residents (including their municipal, First Nations and regional governments) that recognizes that these people have a legitimate stake in the management of the lands the industry wants to use. Successful public engagement starts early in the development process and continues until decommissioning. Its success can be measured by the degree to which the local community provides its informed consent to the development (Herz et al., 2007).

This fifth element of the shale gas development management framework is designed to support a constructive relationship with the population that lives near sites where shale gas exploration and production is conducted. It encompasses two distinct sets of activities: (i) information and consultation; and (ii) good neighbour practices.

9.6.1. Information and Consultation

The process of consultation is not simple, nor is the meaning of consent obvious. In many cases, it is not even obvious who or what constitutes a community; as a consequence, the definition of consent and who can grant it requires careful discussion. However, those discussions must acknowledge the ever-increasing expectations that communities have a say in projects that affect their future.

– Jonathan Lash, President of the World Resources Institute
(Herz et al., 2007)

For any developer, giving residents and local decision-makers timely access to relevant information is a necessary (but insufficient) condition to winning public trust. In the case of shale gas development, such outreach could involve providing information on plans, operations, and performance and include data on the use of water and chemical additives in hydraulic fluids and the production of wastewater and air emissions (IEA, 2012b). An effective public engagement process not only keeps local stakeholders informed through tools such as fact sheets, websites, and open houses, but also establishes a relationship with them (e.g., through an advisory committee) that enables operators to take into account the recommendations of the local residents and process complaints and apply needed corrective measures.

In its discussion paper on a new framework for regulating unconventional oil and gas resources, the AER recommends that operators:

- "Be proactive and consult with landowners, counties, and municipalities to identify opportunities to reduce effects of development
- Provide stakeholders with timely and useful quality information on unconventional resource development.
- "Ensure regionally focused consultation so that stakeholders understand the full scope of development and have clear and fair opportunities to provide input and be familiar with development.
- "Be proactive in increasing local authorities' awareness of any potential activity and effects that may result from development."

(AER, 2012e)

It is clear from this recommendation that effective consultations related to shale gas development will need to take place at different scales (e.g., local and regional), involve different stakeholders or population groups depending on the purpose of the consultation (e.g., residents to discuss minimizing nuisance issues; government planners to discuss water allocation or infrastructure needs), and take place at different stages of the operations cycle (e.g., seek input on exploration plans, report on activities).

9.6.2. Good Neighbour Practices

This set of activities is designed to minimize community disruption during operations. It includes activities such as:

- ensuring public safety;
- minimizing nuisance such as noise, dust and lights;
- reducing to a minimum the area that is affected by the facility;
- respecting the rights of the other stakeholders;
- taking precautions to protect farm and wild animals;
- keeping equipment in good condition;
- driving vehicles safely; and

Management and Mitigation

- reporting damages caused to third parties.

Individual companies and industry associations increasingly recognize the business case for effective public involvement and have developed policies and guides to support this activity (e.g., CAPP, 2003). Several shale gas operators, for example, are already implementing the practices listed in Box 9.3 (Liroff, 2012).

Box 9.3. Documented Good Practices to Gain Community Consent

- Approaches have been identified to help gain community consent for hydraulic fracturing operations. These include:
- Company seeks to secure community consent by initiating contact with local community leaders and organizations and by establishing and implementing a collaborative plan with key stakeholders to identify and address needs and concerns.
- Company has policy relevant to seeking "Free, Prior and Informed Consent" of host communities for new development and activities, such as reaching advance written agreements with local government officials and community organizations outlining company practices related to specific community concerns (noise, setbacks, road use and damage repair, monitoring and addressing social, environmental and health impacts, etc.). Such agreements may include operating practices above and beyond requirements of state regulations and local zoning codes and land use plans applicable to oil and gas drilling and production operations.
- Company has a dedicated hotline to receive individual complaints arising from company operations and has a response tracking mechanism in place to record complaints and company responses.
- Company supports independent third party conflict resolution mechanism to address concerns and complaints arising from company operations in a community.

(Liroff, 2012)

As experience in several countries shows, the manner in which local people are engaged in decisions concerning shale gas development is an important determinant of their acceptance of this development. Residents of shale gas development areas cannot be expected to accept technologies or risks they do not understand. While credible scientific information can help assuage some public concerns, just as important are engagement processes that build relationships and try to reconcile competing perspectives and goals. Failure to consult adequately can be costly and can lead to bad publicity, litigation and, in some cases, even moratoria on development. Such turns of events can be particularly expensive for small companies but can affect the bottom lines of large ones as well (Liroff, 2012).

9.7. CONCLUSION

In this chapter, it has been argued that the protection of public health and the environment during shale gas exploration and production operations requires a systemic approach consisting of (i) sound technologies, (ii) comprehensive management systems, (iii) an effective regulatory system, (iv) the recognition of regional differences, and (v) proactive public engagement. Each of these elements in turn encompasses various activities that together promote process safety, environmental protection, and social acceptance. All of these elements are individually important and the absence of any one weakens the framework's effectiveness, making it vulnerable to unwanted events (see Figure 9.1).

An environmental management framework for shale gas development rests on a solid foundation of environmental monitoring and is supported by five distinct pillars or elements: technology, management systems, regulatory oversight, regional planning, and public engagement. The implementation of such a framework requires a collaborative approach by industry and relevant public authorities.

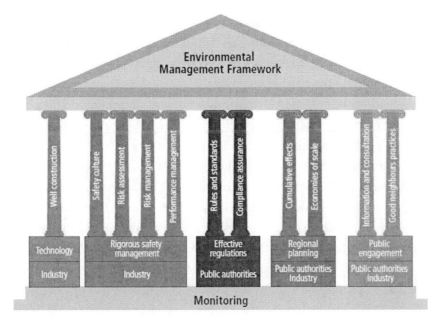

Figure 9.1. Environmental Management Framework. An environmental management framework for shale gas development rests on a solid foundation of environmental monitoring and is supported by five distinct pillars or elements: technology, management systems, regulatory oversight, regional planning, and public engagement. The implementation of such a framework requires a collaborative approach by industry and relevant public authorities.

While most of the elements exist at least in part across the country, important gaps remain in its foundation — in particular, the knowledge required to mitigate the adverse effects of shale gas development on the environment and human health.

No environmental management framework exists in a vacuum and that the ability of organizations to address the multifaceted challenges posed by shale gas development is an important consideration in the framework's successful implementation. An environmental management strategy for shale gas development must do more than identify science gaps and successful mitigation practices. Issues of communication between science and policy, as well as institutional design, performance, and coordination are also fundamental to the successful management of the environmental effects of human activities, such as shale gas development (Young et al., 2008). The challenges of exchanging timely information across large organizations (e.g., governments) and coordinating the actions of various actors with specific mandates (e.g., for process safety, environmental protection, health) are well-known. In the case of shale gas, the need for effective and timely communications among several actors is particularly acute, not least because development is regional rather than local in scale. As well several government organizations have regulatory or policy roles to play, many private operators and their contractors are involved. Moreover, existing residents, including Aboriginal peoples, want to protect their interests.

Organizations will need to be able to adapt to new knowledge about shale gas as it is acquired, and implement new mitigation measures or modify existing ones during the life of development projects. This will require advanced planning to put a systematic process in place for continuous improvement of environmental management practices through learning about their outcomes (an adaptive management approach). The environmental management framework described above is likely to require greater capacity on the part of both operators and regulatory agencies to monitor environmental effects, identify action thresholds in advance and, where required, formulate appropriate management responses.

As widespread public concern about shale gas development indicates, a successful risk-management strategy also involves government authorities and industry winning the public's trust by demonstrating that they apply a comprehensive environmental and safety management system to protect the values the public holds important.

Chapter 10

CONCLUSION

Shale gas represents a vast new energy resource and the world, and its potential environmental impacts are similarly vast. Shale gas is being developed widely, and there are plans for development in China, South America, Europe, Southern Africa, and Australia. Depending on factors such as future natural gas prices and the regulatory environment, further development of shale gas resources will likely span many decades and involve drilling tens of thousands of hydraulically fractured horizontal wells.

The scale and pace at which shale gas resources are being developed are challenging the ability to assess and manage their environmental impacts. The primary concerns are the risk of degradation of the quality of groundwater and surface water (including the safe disposal of large volumes of wastewater; see Chapter 4); the risk of increased GHG emissions (including fugitive methane emissions during and after production), thus exacerbating anthropogenic climate change (see Chapter 5); disruptive effects on communities and the land (see Chapter 6); and adverse effects on human health (see Chapter 7). Other risks include the local release of air contaminants (Chapter 5) and the potential for triggering small to moderate size earthquakes in seismically active areas (Chapter 6). These concerns will vary by region, because of different geological, environmental, and socio-economic conditions, and will depend on the technologies used.

Several tens of thousands of shale gas wells are currently in production. Despite a number of accidents and incidents, the extent and significance of environmental damage is difficult to evaluate because the necessary research and monitoring have not been done. Data are lacking for characterizing and assessing the environmental impacts of shale gas development adequately, particularly in relation to potential groundwater contamination and fugitive methane emissions. There are no vulnerability identification and management systems in place to identify those areas where hydraulic fracturing will be so risky that it should not be undertaken. Although much is known about minimizing the risks related to surface activities, there has been almost no monitoring to assess the risks of gases and fluids contaminating shallow groundwater from below the aquifers as a result of drilling, hydraulic fracturing, inadequate well sealing, and well decommissioning.

These environmental effects are tied with important economic and social issues such as regional economic diversification, energy policy, Aboriginals' rights, and climate change. These broader issues, which lie beyond the scope of this chapter, are nevertheless vital to the conversation on the costs and benefits of shale gas development.

The potential impacts of shale gas development and strategies to manage these impacts need to be considered in the context of global, regional, and local concerns. For example, some communities may welcome shale gas development if it creates jobs and develops the local economy, despite adverse regional or global impacts, whereas others may oppose shale gas because of local impacts, despite regional or global benefits. Moreover, the manner in which local residents are engaged in decisions concerning shale gas development will often influence their acceptance or rejection of this development.

10.1. Summary of the Response to the Charge

What is the state of knowledge of potential environmental impacts from the exploration, extraction, and development of shale gas resources, and what is the state of knowledge of associated mitigation options?

The key findings are as follows:

1. Shale gas development has grown rapidly in the past two decades and many improvements have been made to lessen the potential for environmental impacts including recycling of flowback water, placing more wells per pad, drilling longer laterals resulting in fewer pads and roads, using fewer and more benign chemicals, relying more on tanks rather than ponds to store wastewater, better pond designs, and, in some cases, switching from diesel fuel to natural gas in on-site engines. However, during this time, there has been no comprehensive investment in research and monitoring of environmental and health impacts for either the implementation of best current practices or in the case of accidental releases that cannot be reduced to zero. Many of the pertinent questions are hard to answer objectively and scientifically, either for lack of data, for lack of publicly available data, or due to divergent interpretations of existing data.

2. Natural gas leakage from wells due to improperly formed, damaged or deteriorated cement seals is a long-recognized yet unresolved problem that continues to challenge engineers. Leaky wells are known in some circumstances to create pathways for contamination of groundwater resources and can increase GHG emissions. Conventional methods of monitoring gas leakage may be inaccurate, and are incomplete because leakage outside the main well casing is rarely measured. The issue of well integrity applies to all oil and gas wells, not only shale gas development wells. However, the much larger number of wells needed for shale gas extraction, and the occurrence of shale gas development in areas of

substantial rural and near-urban populations relying on wells for drinking water, suggest that the consequences of leakage will be correspondingly greater than for conventional oil and gas development.

3. An undetermined risk to potable groundwater exists from the upward migration of natural gas and saline waters via complex underground pathways. These pathways include well casing leakage from the production or Intermediate Zones due to inadequate seals, natural fractures in the rock, old abandoned wells, and permeable faults. These connected pathways may allow for migration over the long term, with potentially substantial cumulative impact on aquifer water quality. The monitoring, assessment, and mitigation of impacts from upward migration of contaminants are all more difficult than they are for impacts from surface activities. The potential impacts on groundwater are not currently systematically monitored. Approaches for effective monitoring need to be developed by assessing migration pathways, transport, and attenuation mechanisms and rates specific to site conditions.

4. Gas leakage from wells into freshwater aquifers will not necessarily lead to unacceptable impacts on groundwater. The gas and chemicals formed from reactions of the gas with natural constituents in aquifers may be attenuated close to the wells. However, little is known about assimilation capacity or the resilience of fresh groundwater systems to invasions of stray gas. Investigations specific to these impacts have not been conducted although technology exists for such monitoring. A common claim in the literature is that hydraulic fracturing has shown no verified impacts on groundwater. Recent peer-reviewed literature refutes this claim and also indicates that the main concerns are for longer term cumulative impacts that would generally not yet be evident and are difficult to predict reliably.

5. Fracturing chemicals, flowback water, fuels, and other materials stored temporarily on a well pad during the fracturing stage (generally lasting less than a year), are a potential source of water

Conclusion 201

contamination. About one-quarter to half of the water used in hydraulic fracturing flows back up the well to the surface where it is normally stored in tanks. This flowback is potentially hazardous because it typically contains a portion of the fracturing chemicals, hydrocarbons including benzene and other aromatics, unknown chemicals formed down the well by interactions between chemicals at high temperature and pressure, and constituents leached from the shalesuch as salt, metals, metalloids, and NORM. The flowback is recycled but a residual remains that must be disposed of, either in deep wells where geologically feasible or by treatment and discharge to surface waters. However, suitable, cost-effective treatment is an engineering challenge. Contamination of water from these surface sources can be minimized with appropriate engineering, regulatory enforcement, and performance monitoring.

6. Shale gas development alters the land and local hydrology through the construction of roads, pads, ditches, and pipelines. Although we know a good deal about the effects of road construction, ditches, and pipelines in other contexts, there has not been a comprehensive study of the combined effects specific to shale gas, and some impacts are most likely to be long term. These changes to the land remain until when, decades later, the wells reach a point of diminished production and are abandoned. The degree to which land reclamation will succeed will depend on government policies and economic conditions at the time.

7. Although hydraulic fracturing has caused small earthquakes, seismic risks related to this activity are low, at least in most regions. Hydraulic fracturing near active faults should be avoided and this risk can be diminished by micro-seismic monitoring during operations. Seismic risk caused by the injection of waste fluids can be larger than for hydraulic fracturing, but minimized by careful site selection and monitoring to ensure that excessive injection does not occur.

8. The health risks of shale gas development are not well studied. They include risks to gas field workers and local residents from exposure

to wastewater and air pollution, as well as psycho-social impacts. While shale gas development brings economic benefits, it can also stress community services such as policing, health, and emergency preparedness because of the boomtown effect. Shale gas development can place quality of life and well-being in some communities at risk due to the combination of diverse factors related to the alienation of land, construction of new infrastructure, degradation of water quality, the introduction of nuisances such as truck traffic and noise, loss of rural serenity, and anxiety about unknown impacts. Several of the consequent impacts are expected to be long term.

9. To the extent that natural gas extracted from shale replaces coal in electricity generation, it may also reduce the environmental impact of fossil fuels, including GHG emissions. However, the potential benefits of shale gas over coal are obviated if shale gas displaces low-carbon fuels, such as nuclear energy or renewables including hydro-electric, or if low gas prices discourage investment in efficiency and renewables. These benefits also depend on rates of methane leakage, a subject of continuing inquiry. The net impact of shale gas globally on GHG emissions will thus depend to a significant extent both on control of methane leakage, and on broader energy and climate policies.

10. Shale gas development poses particular challenges for governance because the benefits are primarily regional, though adverse impacts are mostly local and cut across several layers of government. The regulatory framework governing shale gas development is evolving; many aspects are not based on strong science and remain untested. First Nations' rights may be affected in several provinces and need to be considered. Advanced technologies and practices that now exist could minimize many impacts (although some are untested or unverified), but it is not clear that all are economically feasible and that there are technological solutions to address all of the relevant risks. The efficacy of current regulations is not yet established because of the lack of adequate monitoring. The research needed to

Conclusion 203

provide the framework for improved science-based decisions concerning cumulative environmental impacts.

11. Because shale gas development is at an early stage, there is opportunity to implement a variety of measures, including environmental surveillance based on research that will support adaptive approaches to management. Each of the provinces with significant shale gas potential has initiated or is moving toward its own plan for research and monitoring. However there is no national plan and no coordination or federal facilitation of these individual provincial efforts. There can be advantages in the "go slow" approaches allowing additional data collection and integration of multi-disciplinary expertise. There are similar advantages in identifying areas that are too environmentally vulnerable to develop. Given the magnitude of the research needs, strong collaborations involving industry, government, and academia will be necessary. However, achieving public trust in the results will require a high degree of independency of the researchers, transparency, and effective communications.

12. It is evident that more science is needed on which to base regulations, and that such regulations will only be effective if they are informed by timely monitoring and enforced rigorously. Given the current knowledge gaps, a science-based, adaptive, and outcomes-based regulatory approach is more likely to be effective than a prescriptive approach, and is more likely to result in an increase in public trust. The principles of such an approach are well-known and can be found in many existing management systems. Commitment by government and industry to the well-established principles of these management systems will help build public trust.

10.2. A FINAL WORD

The lessons provided by the history of science and technology concerning all major energy sources and many other industrial initiatives show that substantial environmental impacts were typically not anticipated. What is perhaps more alarming is that where substantial adverse impacts were anticipated, these concerns were dismissed or ignored by those who embraced the expected positive benefits of the economic activities that produced those impacts. Many of these adverse impacts could have been lessened, if not entirely avoided, if appropriate management measures, including monitoring programs, had been put in place from the beginning.

Large-scale development of shale gas and, in particular, hydraulic fracturing presents a similar dilemma: promise of significant economic benefits and the possibility of major adverse impacts on people and ecosystems. The highly controversial nature of the topic is manifest in growing concerns expressed by citizens in many parts of the world fueled, in part, by scientific uncertainty as authoritative data about potential impacts are currently neither sufficient, nor conclusive. This seems to be the case both with regard to specific impacts such as the protection of groundwater, as well as to systemic aspects related to broader, long-term questions of energy policy, and community and ecosystem health.

More, well-targeted science is required to ensure that, ultimately, long-term public interests are well understood and safeguarded. Science alone, however, will not address all the relevant concerns because the actual (as opposed to potential) impacts of shale gas development will likely depend to a great extent on the manner in which resource development is managed and regulated. Because shale gas development is still at an early stage, there is opportunity to put in place the management measures required supported by appropriate research to reduce or avoid some of the negative environmental effects of this development. Whether or not shale gas development will turn out in the long term to have been a positive or negative influence on global well-being will depend on how society understands this technology and manages.

GLOSSARY

Additive: Any substance or combination of substances comprised of chemical ingredients found in a hydraulic fracturing fluid, including a proppant, which is added to a base fluid in the context of a hydraulic fracturing treatment.

Annulus: The space surrounding one cylindrical object placed inside another, such as the space in between the casing and the wellbore, or between the casing and tubing, where fluid can flow.

Aquifer: A body of rock that is sufficiently permeable to conduct groundwater and to yield economically significant quantities of water to wells and springs.

Aquitard: A bed of low permeability rock adjacent to an aquifer.

Basin: A closed geologic structure in which the beds dip toward a central location; the youngest rocks are at the center of a basin and are partly or completely ringed by progressively older rocks.

Bedrock: Solid rock either exposed at the surface or situated below surface soil, unconsolidated sediments and weathered rock.

Biocide: An additive that kills bacteria.

Biogenic Methane: As mud turns into shale during shallow burial, generally just a few hundred metres deep, bacteria feed on the available organic matter and release biogenic methane as a byproduct.

Glossary

Blow Out Preventer (BOP): A large valve at the top of the well that may be closed if the drilling crew loses control of formation fluids.

Brine: Water containing salts in solution, such as sodium, calcium, or bromides.

Casing String: Steel piping positioned in a wellbore and cemented in place to prevent the soil or rock from caving in. It also serves to isolate fluids, such as water, gas, and oil, from the surrounding geologic formations.

Casing Shoe: The bottom of the casing string, including the cement around it, or the equipment run at the bottom of the casing string.

Cement Bond Log: A log that uses the variations in amplitude of an acoustic signal traveling down the casing wall to determine the quality of cement bond on the exterior casing wall.

Cement Job: The application of a liquid slurry of cement and water to various points inside or outside the casing.

Centralizer: A device that is used to keep the casing or liner in the center of the wellbore to ensure efficient placement of a cement sheath around the casing string.

Chemical Abstracts Service (CAS): The chemical registry that is the authoritative collection of disclosed chemical substance information.

Chemical Ingredient: A discrete chemical constituent with its own specific name or identity, such as a CAS number, that is contained in an additive.

Communication: The flow of fluids from one part of a reservoir to another or from the reservoir to the wellbore.

Completion: The activities and methods to prepare a well for production and following drilling. Includes installation of equipment for production from a gas well.

Corrosion Inhibitor: A chemical additive used to protect iron and steel components in the wellbore and treating equipment from the corrosive treating fluid.

Disposal Well: A well which injects produced water into an underground formation for disposal.

Domestic Water Well: An opening in the ground, whether drilled or altered from its natural state, for the production of groundwater used for drinking, cooking, washing, yard, or livestock use.

Glossary

Flowback: The process of allowing fluids to flow from the well following a treatment, either in preparation for a subsequent phase of treatment or in preparation for cleanup and returning the well to production.

Formation (Geologic): A rock body distinguishable from other rock bodies and useful for mapping or description. Formations may be combined into groups or subdivided into members.

Formation Fluids: Any fluid that occurs in the pores of a rock.

Fractures Networks: Patterns in multiple fractures that intersect each other.

Fracturing Fluids: The fluid used to hydraulically induce cracks in the target formation and includes the applicable base fluid and all additives.

Fresh (Non-Saline) Groundwater: Groundwater that has a total dissolved solids (TDS) content less than or equal to 4,000 mg/L or as defined by regulation.

Gas Migration: A flow of gas that is detectable at surface outside of the outermost casing string. It refers to all possible routes for annular gas entry and propagation through and around the cement sheath.

Groundwater: Subsurface water that is in the zone of saturation; source of water for wells, seepage, and springs. The top surface of the groundwater is the water table.

Horizontal Drilling: A drilling procedure in which the wellbore is drilled vertically to a kickoff depth above the target formation and then angled through a wide 90 degree arc such that the producing portion of the well extends horizontally through the target formation.

Hydraulic Fracturing: Injecting fracturing fluids into the target formation at a force exceeding the parting pressure of the rock thus inducing a network of fractures through which oil or natural gas can flow to the wellbore.

Injection Well: A well-used to inject fluids into an underground formation either for enhanced recovery or disposal.

Intermediate Casing: A casing string that is generally set in place after the surface casing and before the production casing to provide protection against caving of weak or abnormally pressured formations.

Kerogen: The naturally occurring, solid, insoluble organic matter that occurs in source rocks and can yield oil upon heating. Kerogens have a high molecular weight relative to bitumen, or soluble organic matter. Bitumen forms from kerogen during petroleum generation. Kerogens are described as Type I, consisting of mainly algal and amorphous (but presumably algal) kerogen and highly likely to generate oil; Type II, mixed terrestrial and marine source material that can generate waxy oil; and Type III, woody terrestrial source material that typically generates gas.

Original Gas-in-Place: The entire volume of gas contained in the reservoir, regardless of the ability to produce it.

Pad: Cleared ground surface, usually covered in gravel, used to drill a well and store equipment.

Permeability: A rock's capacity to transmit a fluid. A rock may have significant porosity (many microscopic pores) but have low permeability if the pores are not interconnected. Permeability may also exist or be enhanced through fractures that connect the pores.

Play: A conceptual model for a style of hydrocarbon accumulation used by oil and gas companies to develop prospects in a basin, region or trend.

Porosity: The percentage of pore volume or void space or that volume within rock that can contain fluids.

Produced Water: Water naturally present in a reservoir or injected into a reservoir to enhance production, produced as a co-product when gas or oil is produced.

Production Casing: A casing string that is set across the reservoir interval and within which the primary completion components are installed.

Propping Agents/Proppant: Synthetic or natural non-compressible grains such as coated sand or sintered bauxite ceramics pumped into a formation during a hydraulic fracturing operation to hold fractures open around the wellbore and to enhance fluid extraction after hydraulic fracturing pressures are removed.

Recoverable Resources: The volume of resource that is technically or economically feasible to extract.

Recycle: The process of treating flowback or produced water to allow it to be reused either for hydraulic fracturing or for another purpose.

Reuse: The process of using water multiple times for similar purposes.

Glossary

Saline Groundwater: Groundwater that has a total dissolved solids (TDS) content more than 4,000 mg/L or as defined by the jurisdiction.

Shale Gas: Natural gas produced from low permeability shale formations.

Slickwater: A water based fluid mixed with friction reducing agents, commonly potassium chloride.

Stimulation: Any of several processes used to enhance near wellbore permeability and reservoir permeability. Stimulation in shale gas reservoirs typically takes the form of hydraulic fracturing treatments.

Stray or Fugitive Gas: Terms usually applied to gas that has escaped from a well or gas-handling facility near the oil or gas well and has been detected in a location where it is unwanted (e.g., gas that has migrated into shallow groundwater from a geological source most likely through a poorly or incompletely cemented well).

Surface Casing: A large diameter, relatively low pressure pipe string set in shallow yet competent formations to protect fresh water aquifers onshore, provide minimum pressure integrity, and enables a diverter or blow out preventer to be attached to the top of the string. It also provides structural strength so that the remaining casing may be suspended at the top and inside the casing.

Surface Water: Water collecting on the ground or in a stream, river, lake, sea, or ocean, as opposed to groundwater.

Technically Recoverable Resources: The total amount of resource, discovered and undiscovered, that is thought to be recoverable with available technology, regardless of economics.

Thermogenic Methane: Natural gas generated during deep burial, generally several kilometres deep, where heat and pressure crack the organic matter, including any oil already produced by the same heat and pressure, into smaller hydrocarbons.

Tight Gas: Natural gas trapped in a hardrock, sandstone, or limestone formation that is relatively impermeable.

Total Organic Carbon: The concentration of organic material in source rocks as represented by the weight percent of organic carbon. A value of 2 percent is considered the minimum for shale gas reservoirs.

Unconventional: Oil and gas resources whose porosity, permeability, fluid trapping mechanism, or other characteristics differ from conventional sandstone and carbonate reservoirs.

Vadose Zone: Ground between the land surface and the top of the water table.

Wait on Cement (WOC): To suspend drilling operations while allowing cement slurries to solidify, harden and develop compressive strength.

Wastewater: Spent or used water with dissolved or suspended solids, discharged from homes, commercial establishments, farms, and industries.

Wellbore: A wellbore is the open hole that is drilled prior to the installation of casing and cement.

Wet Gas: Natural gas with liquid hydrocarbon or condensate phases present Taken from (ALL Consulting, 2012; Jackson et al., 2013b)

REFERENCES

ACOLA (Australian Council of Learned Academies). (2013). *Engineering Energy: Unconventional Gas Production*. Melbourne, Australia: ACOLA.

Adams, C. (2012). *Summary of Shale Gas Activity in Northeast British Columbia 2011*. Victoria (BC): B.C. Ministry of Energy and Mines.

Adams, M. B. (2011). Land application of hydrofracturing fluids damage a deciduous forest stand in West Virgina. *Journal of Environmental Quality, 40*, 1340-1344. doi: 10.2134/jeq2010.0504.

AEMP (Alberta Environmental Monitoring Panel). (2011). *A World Class Environmental Monitoring, Evaluation and Reporting System for Alberta*. Edmonton (AB): AEMP.

Al, T., Butler, K., Cunjak, R., & MacQuarrie, K. (2012). *Opinion: Potential Impact of Shale Gas Exploitation on Water Resources*. Fredericton (NB): University of New Brunswick.

Allen, D. T., Torres, V. M., Thomas, J., Sullivan, D. W., Harrison, M., Hendler, A., Seinfeld, J. H. (2013). Measurements of methane emissions at natural gas production sites in the United States. *Proceeding of the National Academy of Sciences. 44* (4). doi: 10.1073/pnas.1304880110.

Allen, M. R., Frame, D. J., Huntingford, C., Jones, C. D., Lowe, J. A., Meinshausen, M., & Meinshausen, N. (2009). Warming caused by cumulative carbon emissions towards the trillionth tonne. *Nature, 458*(7242), 1163-1166. doi: 10.1038/nature08019.

Aller, L., Bennett, T., Lehr, J. H., Petty, R. J., & Hackett, G. (1987). *DRASTIC: A Standardized System for Evaluating Ground Water Pollution Potential Using Hydrogeologic Settings.* Dublin (OH): U.S. EPA.

Alvarez, R. A., Pacala, S. W., Winebrake, J. J., Chameides, W. L., & Hamburg, S. P. (2012). Greater focus needed on methane leakage from natural gas infrastructure. *Proceedings of the National Academy of Sciences of the United States of America, 109*(17), 6435-6440. doi: 10.1073/pnas.1202407109.

Amos, R. T. & Blowes, D. W. (2008). Versatile direct push profiler for the investigation of volatile compounds near the water table. *Water Resources Research, 44*(4). doi: 10.1029/2008WR006936.

ANGA (America's Natural Gas Alliance) & AXPC (American Exploration and Production Council). (2012). *ANGA Comments to EPA on New Source Performance Standards for Hazardous Air Pollutants Review America's Natural Gas Alliance.* Washington (DC): ANGA and AXPC.

APHA (American Public Health Association). (2012). *Policy Statement on the Environmental and Occupational Health Impacts of High-Volume Hydraulic Fracturing of Unconventional Gas Reserves.* Washington (DC): APHA.

API (American Petroleum Institute). (2009). *Hydraulic Fracturing Operation – Well Construction and Integrity Guidelines.* Washington (DC): API.

ARI (Advanced Resources International, Inc.). (2013). *EIA/ARI World Shale Gas and Shale Oil Resource Assessment. Technically Recoverable Shale Gas and Shale Oil Resources: An Assessment of 137 Shale Formations in 41 Countries Outside the United States.* Arlington(VA): ARI.

Arthur, J. D., Bohm, B., Coughlin, B. J., & Layne, M. (2008). *Evaluating the Environmental Implications of Hydraulic Fracturing in Shale Gas Reservoirs.* Tulsa (OK): ALL Consulting.

Arthur, J. D., Coughlin, B. J., Bohm, B. K., & ALL Consulting. (2010). *Summary of Environmental Issues, Mitigation Strategies, and*

References

Regulatory Challenges Associated with Shale Gas Development in the United States and Applicability to Development and Operations in Canada. Paper presented at Canadian Unconventional Resources and International Petroleum Conference, Calgary (AB).

AWWA (American Water Works Association). (2013). *Water and Hydraulic Fracturing.* Denver (CO): AWWA. B.C. Ministry of Energy and Mines & NEB (National Energy Board). (2011). *Ultimate Potential for Unconventional Natural Gas in Northeastern British Columbia's Horn River Basin.* Victoria (BC): B.C. Ministry of Energy and Mines and NEB.

Bamberg, S. & Möser, G. (2007). Twenty years after Hines, Hungerford, and Tomera: A new meta-analysis of psycho-social determinants of pro-environmental behaviour. *Journal of Environmental Psychology, 27,* 14-25.

Bamberger, M. & Oswald, R. E. (2012). Impacts of gas drilling on human and animal health. *New Solutions, 22*(1), 51-77.

BAPE (Bureau d'audiences publiques sur l'environnement). (2011b). *Développement durable de l'industrie des gaz de schiste au Québec.* [*Sustainable Development of the Shale Gas Industry in Québec.*] *Rapport 273.* Québec (QC): BAPE.

Barker, J. F. & Fritz, P. (1981). The occurrence and origin of methane in some groundwater flow systems. *Canadian Journal of Earth Sciences, 18*(12), 1802-1816. doi: 10.1139/e81-168.

Birkholzer, J. T., Nicot, J. P., Oldenburg, C. M., Zhou, Q., Kraemer, S., & Bandilla, K. (2011). Brine flow up a well caused by pressure perturbation from geologic carbon sequestration: Static and dynamic evaluations. *International Journal of Greenhouse Gas Control, 5*(4), 850-861. doi: 10.1016/j. ijggc.2011.01.003.

Bjørlykke, K. (1989). *Sedimentology and Petroleum Geology.* Berlin, Germany: Springer-Verlag.

Bloomberg, M. R. & Mitchell, G. P. (2012, August 23). Fracking is too Important to Foul Up, *Washington Post.*

214 *References*

Bourgoyne, A. T., Scott, S. L., & Manowski, W. (2000). *A Review of the Sustained Casing Pressure Occurring on the OCS.* Washington (DC): U.S. Department of the Interior, Minerals Management Services.

Bradbury, J., Obeiter, M., Draucker, L., Wang, W., & Stevens, A. (2013). *Clearing the Air: Reducing Upstream Greenhouse Gas Emissions from U.S. Natural Gas Systems.* Working Paper. Washington (DC): World Resources Institute.

Broomfield, M. (2012). *Support to the Identification of Potential Risks for the Environment and Human Health Arising from Hydrocarbon Operations Involving Hydraulic Fracturing in Europe.* Didcot, United Kingdom: European Commission DG Environment.

Brouyère, S., Jeannin, P. Y., Dessargues, A., Goldscheider, N., Popescu, I. C., Sauter, M., Zwahlen, F. (2011). Evaluation and Validation of Vulnerability Concepts Using a Physically Based Approach. Paper presented at 7th *Conference on Limestone Hydrology and Fissure Media*, Besancon, France.

Brown, A. (2000). Evaluation of possible gas microseepage mechanisms. *American Association of Petroleum Geologists Bulletin, 84*(11), 1775-1789.

Brown, J. (2010). *Drilling 101.* Paper presented in person, Sibley Hall, Cornell University, Ithica (NY).

Burnham, A., Han, J., Clark, C. E., Wang, M., Dunn, J. B., & Palou-Rivera, I. (2012). Life-cycle greenhouse gas emissions of shale gas, natural gas, coal, and petroleum. *Environmental Science and Technology, 46*(2), 619-627.

Caboit Oil & Gas Corporation. (2012). *Uniquely Cabot. 2012 Annual Report.* Houston (TX): Caboit Oil & Gas Corporation, Canadian Water Network. (2013). Hydraulic Fracturing and Water Competition Announcement. Retrieved January 2014, from http://www.cwn-rce.ca/initiatives/research-calls/hydraulic-fracturing-and-water-competition-announcement/.

CAPP (Canadian Association of Petroleum Producers). (2003). *Guide for Effective Public Involvement.* Calgary (AB): CAPP.

References

CAPP (Canadian Association of Petroleum Producers). (2012a). *Responsible Canadian Energy. 2012 Progress Report Summary.* Calgary (AB): CAPP.

CAPP (Canadian Association of Petroleum Producers). (2012b). *Baseline Groundwater Testing.* Calgary (AB): CAPP.

CAPP (Canadian Association of Petroleum Producers). (2012c). *Anomalous Induced Seismicity: Assessment, Monitoring, Mitigation and Response.* Calgary (AB): CAPP.

CAPP (Canadian Association of Petroleum Producers). (2012d). *Fluid Transport, Handling, Storage and Disposal.* Calgary (AB): CAPP.

CAPP (Canadian Association of Petroleum Producers). (2012e). *Water Sourcing, Measurement, and Reuse.* Calgary (AB): CAPP.

CAPP (Canadian Association of Petroleum Producers). (2012f). *Wellbore Construction and Quality Assurance.* Calgary (AB): CAPP.

CAPP (Canadian Association of Petroleum Producers). (2012g). *Fracturing Fluid Additive Risk Assessment and Management.* Calgary (AB): CAPP.

CAPP (Canadian Association of Petroleum Producers). (2012h). *Fracturing Fluid Additive Disclosure.* Calgary (AB): CAPP.

CAPP (Canadian Association of Petroleum Producers). (2014). *Basic Statistics.* Retrieved January 2014, from http://www.capp.ca/library/statistics/basic/Pages/default.aspx.

Carter, T., Fortner, L., & Béland Otis, C. (2009). *Shale Gas Opportunities in Southern Ontario.* Paper presented at 48th Annual OPI Conference and Trade Show, Sarnia (ON).

Cathles, L. M., Brown, L., Taam, M., & Hunter., A. (2012). A commentary on "The greenhouse-gas footprint of natural gas in shale formations" by R. W. Howarth, R. Santoro, and Anthony Ingraffea. *Climate Change, 113*(2), 525–535.

CDPHE (Colorado Department of Public Health and Environment). (2010). *Public Health Implications of Ambient Air Exposures as Measured in Rural and Urban Oil & Gas Development Areas – An Analysis of 2008 Air Sampling Data.* Denver (CO): CDPHE.

Chapman, S. W., Parker, B. L., Cherry, J. A., McDonald, S. D., Goldstein, K. J., Frederick, J. J.,... Williams, C. E. (2013). Combined MODFLOW-

FRACTRAN application to assess chlorinated solvent transport and remediation in fractured sedimentary rock. *Remediation, 23*(3), 7-35.

Chesapeake Energy. (2012). *Water Used in Deep Shale Gas Exploration Fact Sheet.* Oklahoma City (OK): Chesapeake Energy.

Cheung, K., Klassen, P., Mayer, B., Goodarzi, F., & Aravena, R. (2010). Major ion and isotope geochemistry of fluids and gases from coalbed methane and shallow groundwater wells in Alberta, Canada. *Applied Geochemistry, 25*, 1307-1329.

Clark, I. D., Al, T., Jensen, M., Kennell, L., Mazurek, M., Mohapatra, R., & Raven, K. G. (2013). Paleozoic-aged brine and authigenic helium preserved in an Ordovician shale aquiclude. *Geology, 41*(9), 951-954.

Cohen, H. A., Parratt, T., & Andrews, C. B. (2013). Discussion of papers "Potential contaminant pathways from hydraulically fractured shale to aquifers," *Groundwater, 51*(3), 317-319.

Colborn, T., Kwiatkowski, C., Schultz, K., & Bachran, M. (2011). Natural gas operations from a public health perspective. *Human and Ecological Risk Assessment, 17*, 1039-1056.

Comité de l'évaluation environnementale stratégique sur le gaz de schiste. (2012). *Implementation Plan for the Strategic Environmental Assessment on Shale Gas.* Québec (QC): Gouvernement du Québec.

Committee of Energy and Commerce. (2011). *Chemicals Used in Hydraulic Fracturing.* Washington (DC): U.S. House of Representatives.

Cooke Jr., C. E. (1979). Radial differential temperature (RDT) logging – A new tool for detecting and treating flow behind casing. *Journal of Petroleum Technology, 31*(6), 676-682.

Cooley, H. & Donnelly, K. (2012). *Hydraulic Fracturing and Water Resources: Separating the Frack from the Fiction.* Oakland (CA): Pacific Institute.

Cottle, M. & Guidotti, T. (1990). Process chemical in the oil and gas industry: Potential occupational hazards, toxicology, and industrial health. *Toxicology and Industrial Health, 6*, 41-56.

Coussens, C. & Martinez, R. M. (2014). *Health Impact Assessment of Shale Gas Extraction: Workshop Summary.* Washington (DC): National Academy of Sciences.

References

217

Covello, V. T. (1983). The perception of technological risks: A literature review. *Technological Forecasting and Social Change, 23*(4), 285-297.

Covello, V. T. (1992). Risk Communication: An Emerging Area of Health Communication Research. In S. A. Deetz (Ed.), *Communications Yearbook.* Newbury Park (CA): Sage Publications.

Crouse, D. L., Peters, P. A., Van Donkelaar, A., Goldberg, M. S., Villeneuve, P. J., Brion, O., Burnett, R. T. (2012). Risk of nonaccidental and cardiovascular mortality in relation to long-term exposure to low concentrations of fine particulate matter: A Canadian national-level cohort study. *Environmental Health Perspectives, 120*(5), 708-714.

CSA Group (Canadian Standards Association Group). (2012). *Geological Storage of Carbon Dioxide (Z741-12).* Toronto (ON): CSA Group.

Daigle, P. (2010). A summary of the environmental impacts of roads, management responses, and research gaps: A literature review. *BC Journal of Ecosystems and Management, 10*(3), 65-89.

Davey, E. (2012). Written Ministerial Statement by Edward Davey: *Exploration for Shale Gas.* Retrieved January 2014, from https://www.gov.uk/government/speeches/written-ministerial-statement-by-edward-davey-exploration-for-shale-gas.

DNV (Det Norske Veritas). (2013). *Recommended Practice: Risk Management of Shale Gas Developments and Operations (DNV-RP-U301).* Høvik, Norway: DNV.

Dusseault, M., Gray, M., & Nawrocki, P. (2000). Why Oilwells Leak: Cement Behavior and Long-Term Consequences. Paper presented at *Society of Petroleum Engineers International Oil and Gas Conference and Exhibition*, Beijing, China.

Dusseault, M. (2013). Geomechanical Aspects of Shale Gas Development. Paper presented at *EUROCK 2013*, Wroclaw, Poland.

EC/R Incorporated. (2011). *Oil and Natural Gas Sector: Standards of Performance for Crude Oil and Natural Gas Production, Transmission, and Distribution.* Background Technical Support Document for Proposed Standards. Chapel Hill (NC): EPA.

EEA (European Environmental Agency). (2001). *Late Lessons from Early Warnings: The Precautionary Principle 1896–2000*. Copenhagen, Denmark: EEA.

EEA (European Environmental Agency). (2013). *Late Lessons from Early Warnings: Science, Precaution, Innovation*. Copenhagen, Denmark: EEA.

Efstathiou, J. & Drajem, M. (2013). *Drillers Silence Fracking Claims With Sealed Settlements*. Retrieved August 2013, from http://www.bloomberg.com/news/2013-06-06/drillers-silence-fracking-claims-with-sealed-settlements.html.

EIA (U.S. Energy Information Administration). (1998). *Natural Gas 1998. Issues and Trends*. Washington (DC): EIA.

EIA (U.S. Energy Information Administration). (2011). Maps: Exploration, Resources, Reserves, and Production. Retrieved August 2013, from http://www.eia.gov/pub/oil_gas/natural_gas/analysis_publications/maps/maps.htm.

EIA (U.S. Energy Information Administration). (2012). *Annual Energy Outlook 2012*. Washington (DC): EIA.

EIA (U.S. Energy Information Administration). (2014). *Henry Hub Natural Gas Spot Price*. Retrieved January 2014, from http://www.eia.gov/dnav/ng/hist/rngwhhdm.htm.

Einarson, M. (2006). Multilevel Groundwater Monitoring. In *Practical Handbook of Environmental Site Characterization and Groundwater Monitoring* (2nd edition). Boca Raton (FL): CRC Taylor & Francis.

Einarson, M. D. & Cherry, J. A. (2002). A new multilevel ground water monitoring system using multichannel tubing. *Groundwater Monitoring and Remediation, 22*(4), 52-65.

Eklund, D. (2005). *Penetration Due to Filtration Tendency of Cement Based Grouts*. Stockholm, Sweden: Royal Institute of Technology.

Eligon, J. (2013, January 27). An Oil Boom Takes a Toll on Health Care, *The New York Times*.

Encana. (2013). Courtesy Matters. Retrieved September 2013, from http://www.encana.com/communities/courtesy-matters/.

References 219

Entrekin, S., Evans-White, M., Johnson, B., & Hagenbuch, E. (2011). Rapid expansion of natural gas development poses a threat to surface waters. *Frontiers in Ecology and the Environment, 9*(9), 503-511. doi: 10.1890/110053. Environment Canada. (2011). *An Integrated Oil Sands Environment Monitoring.*

EPA (Environmental Protection Agency). (1993). *Guidance for Evaluation the Technical Impracticability of Ground-Water Restoration.* Washington (DC): EPA.

EPA (Environmental Protection Agency). (2009a). *Greenhouse Gas Emissions Reporting From the Petroleum and Natural Gas Industry. Background Technical Support Document.* Washington (DC): EPA.

EPA (Environmental Protection Agency). (2009b). *National Primary Drinking Water Regulations.* Washington (DC): EPA.

EPA (Environmental Protection Agency). (2009c). *Ozone and Your Health.* Washington (DC): EPA.

EPA (Environmental Protection Agency). (2010). *Inventory of U.S. Greenhouse Gas Emissions and Sinks: 1990–2008.* Washington (DC): EPA.

EPA (Environmental Protection Agency). (2011a). *Oil and Natural Gas Sector: New Source Performance Standards and National Emissions Standards for Hazardous Air Pollutants Reviews.* Washington (DC): EPA.

EPA (Environmental Protection Agency). (2011b). *Draft Plan to Study the Potential Impacts of Hydraulic Fracturing on Drinking Water Resources.* Washington (DC): EPA.

EPA (Environmental Protection Agency). (2012a). *Regulation of Hydraulic Fracturing Under the Safe Drinking Water Act.* Retrieved July 2013, from http://water.epa.gov/type/groundwater/uic/class2/hydraulicfract uring/wells_hydroreg.cfm.

EPA (Environmental Protection Agency). (2012b). *Health Effects.* Retrieved September 2013, from http://www.epa.gov/glo/health.html.

EPA (Environmental Protection Agency). (2012c). *Oil and Natural Gas Sector: Standards of Performance for Crude Oil and Natural Gas Production, Transmission, and Distribution. Background Supplemental*

Technical Support Document for the Final New Source Performance Standards. Chapel Hill (NC): EPA.

EPA (Environmental Protection Agency). (2013a). *Nitrogen Dioxide. Health.* Retrieved September 2013, from http://www.epa.gov/airquality/nitrogenoxides/health.html.

EPA (Environmental Protection Agency). (2013b). *EPA Needs to Improve Air Emissions Data for the Oil and Natural Gas Production Sectors.* Washington (DC): EPA.

EPA (Environmental Protection Agency). (2013c). *Draft Inventory of U.S. Greenhouse Gas Emissions and Sinks: 1990–2011.* Washington (DC): EPA.

EPA (Environmental Protection Agency). (2013d). P*articulate Matter (PM). Health.* Retrieved September 2013, from http://www.epa.gov/pm/health.html.

Erickson, P. A. (1994). *A Practical Guide to Environmental Impact Assessment.* San Diego (CA): Academic Press.

Erno, B. & Schmitz, R. (1996). Measurements of soil gas migration around oil and gas wells in the Lloydminster area. *Journal of Canadian Petroleum Technology, 35*(7). doi: 10.2118/96-07-05.

Ethridge, S. (2010). *Interoffice Memorandum: Health Effects Reviews of Barnett Shale Formation Area Monitoring Projects.* Austin (TX): Texas Commission on Environmental Quality.

European Parliament. (2011). *Impacts of Shale Gas and Shale Oil Extraction on the Environment and on Human Health.* Brussels, Belgium: European Parliament.

Ewen, C., Borchardt, D., Richter, S., & Hammerbacher, R. (2012). H*ydrofracking Risk Assessment. Study Concerning the Safety and Environmental Compatibility of Hydrofracking for Natural Gas Production from Unconventional Reservoirs (Executive Summary).* Berlin, Germany: ExxonMobil Production Deutschland Gmbh.

Fierro, M. A., O'Rourke, M. K., & Burgess, J. L. (2001). *Adverse Health Effects of Exposure to Ambient Carbon Monoxide.* Tucson (AZ): College of Public Health, University of Arizona.

References 221

Fisher, K. & Warpinski, N. (2011). *Hydraulic Fracture-Height Growth: Real Data.* Paper presented at Society of Petroleum Egineers Annual Technical Conference and Exhibition, Denver (CO).

Flemisch, B., Darcis, M., Erbertseder, K., Faigle, B., Lausser, A., Mosthaf, K., Helmig, R. (2011). DuMu^x: Dune for multi-flow and transport in porous media. *Advances in Water Resources, 34*, 1102-1112.

Flewelling, S. A., Tymchak, M. P., & Warpinski, N. (2013). Hydraulic fracture height limits and fault interactions in tight oil and gas formations. *Geophysical Research Letters 40*, 1-5. doi: 10.1002/grl.50707.

Flewelling, S. A. & Sharma, M. (2014). Constraints on upward migration of hydraulic fracturing fluid and brine. *Groundwater, 52*(1). doi: 10.1111/ gwat.12095.

FNFN (Fort Nelson First Nation). (2012). *Respect for the Land: Fort Nelson First Nation Strategic Land Use Plan.* Fort Nelson (BC): Fort Nelson First Nation.

Focazio, M. J., Reilly, T. E., Rupert, M. G., & Helsel, D. R. (2002). *Assessing Ground-Water Vulnerability to Contamination: Providing Scientifically Defensible Information for Decision Makers.* Denver (CO): USGS.

Folkes, D. J. (1982). Control of contaminant migration by the use of liners. *Canadian Geotechnical Journal 19*(3), 320-344.

Fountain, J. C. & Jacobi, R. D. (2000). Detection of buried faults and fractures using soil gas analysis. *Environmental and Engineering Geoscience, 6*(3), 201-208.

Fraser Basin Council. (2012). *Identifying Health Concerns Relating to Oil & Gas Development in Northeastern BC.* Vancouver (BC): BC Ministry of Health.

Frind, E. O., Molson, J. W., & Rudolph, D. L. (2006). Well vulnerability: A quantitative approach for source water protection. *Groundwater, 44*(5), 732-742. doi: 10.1111/j.1745-6584.2006.00230.x.

Gassiat, C., Gleeson, T., Lefebvre, R., & McKenzie, J. (2013). Hydraulic fracturing in faulted sedimentary basins: Numerical simulation of potential contamination of shallow aquifers over long time scales. *Water*

Resources Research, 40(12), 8310-8327. doi: 10.1002/2013WR014287.

Gautschi, A. (2001). Hydrogeology of a fractured shale (Opalinus Clay): Implications for deep geological disposal of radioactive wastes. *Hydrogeology Journal, 9,* 97-107.

GE (General Electric). (2014). *FlexEfficiency* 50 Combined Cycle Power Plant.* Retrieved January 2014, from http://www.ge-energy.com/products_and_services/products/gas_turbines_heavy_duty/flexefficiency_50_combined_cycle_power_plant.jsp.

Geoscience B.C. (2011a). *Geoscience BC Announces Collaborative Horn River Basin Water Study Phase II Activities.* Retrieved January 2014, from http://www.geosciencebc.com/s/NewsReleases.asp?ReportID=478074.

Geoscience B.C. (2011b). *Montney Water Project. Project Completion Report.* Vancouver (BC): Geoscience B.C.

Geoscience B.C. (2012). *Geoscience BC Announces Collaborative Regional Seismic Program.* Retrieved April 2013, from http://www.geosciencebc.com/s/NewsReleases.asp?ReportID=545816&_Type=News&_Title=Geoscience-BC-Announces-Collaborative-Regional-Seismic-Program.

Gevantman, L. H. (2013). Solubility of Selected Gases. In W. M. Haynes (94th Ed.), *Handbook of Chemistry and Physics.* Boca Raton (FL): CRC Press/ Taylor and Francis.

Gilman, J. B., Lerner, B. M., Kuster, W. C., & de Gouw, J. A. (2013). Source signature of volatile organic compounds from oil and natural gas operations in northeastern Colorado. *Environmental Science and Technology, 47*(3), 1297-1305. doi: 10.1021/es304119a.

Gimmi, T., Waber, H. N., Gautschi, A., & Rubel, A. (2007). Stable water isotopes in pore water of Jurassic argillaceous rocks as tracers for solute transport over large spatial and temporal scales. *Water Resources Research, 43*(4). doi: 10.1029/2005WR004774.

Goldstein, B. D., Kriesky, J., & Pavliakova, B. (2012). Missing from the table: Role of the environmental public health community in governmental advisory commissions related to Marcellus Shale drilling. *Environmental Health Perspectives, 120*(4), 483-486.

Goldstein, B. D., Bjerke, E. F., & Kriesky, J. (2013). Challenges of unconventional shale gas development: So what's the rush?. *Notre Dame Journal of Law, Ethics & Public Policy, 27*(1), 149-185.

GoodGuide. (2011). *Hazardous Air Pollutants (Clean Air Act).* Retrieved January 2014, from http://scorecard.goodguide.com/chemical-groups/one-list.tcl?short_list_name=hap.

Gorody, A. W. (2012). Factors affecting the variability of stray gas concentration and composition in groundwater. *Environmental Geosciences, 19*(1), 17-31. doi: 10.1306/eg.12081111013.

Gregory, K. B., Radisav, D. V., & Dzombak, D. A. (2011). Water management challenges associated with the production of shale gas by hydraulic fracturing. *Elements, 7*(3), 181-186. doi: 10.2113/gselements.7.3.181.

GRES (Groupe de recherche sur l'eau souterraine). (2013). Projet/PACES/Mise en Contexte. [Project / PACES / Background.] Retrieved January 2014, from http://gres.uqat.ca/FR/PACES_MISE_EN_CONTEXTE.

GW Solutions. (2012). *Montney Water Project – Hydrogeologic Review.* Dawson Creek (BC): City of Dawson Creek.

Halliburton. (2008). *U.S. Shale Gas. An Unconventional Resource. Unconventional Challenges.* Houston (TX): Halliburton.

Hamlat, M. S., Djeffal, S., & Kadi, H. (2001). Assessment of radiation exposures from naturally occurring radioactive materials in the oil and gas industry. *Applied Radiation and Isotopes, 55*(1), 141-146.

Hawkes, C. D., McLellan, P. J., Zimmer, U., & Bachu, S. (2004). Geomechanical Factors Affecting Geological Storage of CO_2 in Depleted Oil and Gas Reservoirs: Risks and Mechanisms. Paper presented at *Gulf Rocks 2004, the 6th North America Rock Mechanics Symposium (NARMS)*, Houston (TX).

Hayhoe K., Kheshgi H. S., Jain A. K., & Wuebbles D. J. (2002). Substitution of natural gas for coal: Climatic effects of utility sector emissions. *Climate Change, 54*, 107-139.

Healy, J. H., Rubey, W. W., Griggs, D. T., & Raleigh, C. B. (1968). The Denver earthquakes. *Science, 161*(3848), 1301-1310.

Heath, G. A. & Mann, M. K. (2012). Background and reflections on the life cycle assessment harmonization project. *Journal of Industrial Ecology, 16*(S1), S8-S11.

Heilweil, V. M., Stolp, B. J., Kimball, B. A., Susong, D. D., Marston, T. M., & Gardner, P. M. (2013). A stream-based methane monitoring approach for evaluating groundwater impacts associated with unconventional gas development. *Groundwater, 51*(4), 511-524.

Herz, S., Vina, A., & Sohn, J. (2007). *Development without Conflict. The Business Case for Community Consent.* Washington (DC): World Resources Institute.

Hill, J. (2013). *Eagle Ford Trucking Accident Risk.* Retrieved August 2013, from http://www.rhlawgroup.com/blog/eagle-ford-trucking-accident-risk/.

Holder, J. (2004). *Environmental Assessment: The Regulation of Decision Making.* New York (NY): Oxford University Press.

Horn River Basin Producers Group. (2010). *Horn River Basin/Shale Gas Frequently Asked Questions.* Retrieved September 2013, from http://www.northernrockies.ca/assets/Business/Producers~Group/HRB PG_FAQ.pdf.

House of Commons Energy and Climate Change Committee. (2011). *Shale Gas. Fifth Report of Sessions 2010–12.* London, United Kingdom: House of Commons.

Howarth, R., Shindell, D., Santoro, R., Ingraffea, A., Phillips, N., & Townsend-Small, A. (2012). *Methane Emissions from Natural Gas Systems.* Washington (DC): National Climate Assessment.

Howarth, R., Santoro, R., & Ingraffea, A. (2011). Methane and the greenhouse-gas footprint of natural gas from shale formations. *Climate Change, 106*(4), 670-690. doi: 10.1007/s10584-011-0061-5.

Hughes, D. J. (2013). *Drill, Baby, Drill.* Santa Rosa (CA): Post Carbon Institute. Hughes, T. P. (1983). *Networks of Power: Electrification in Western Society, 1880–1930.* Baltimore (MD): Johns Hopkins University Press.

References 225

Hultman, N., Rebois, D., Scholten, M., & Ramig, C. (2011). The greenhouse impact of unconventional gas for electricity generation. *Environmental Research Letters, 6*(4). doi: 10.1088/1748-9326/6/4/044008.

IAIA (International Association for Impact Assessment). (1999). *Principles of Environmental Impact Assessment Best Practice.* Fargo (ND): IAIA.

IEA (International Energy Agency). (2011). *Are We Entering a Golden Age of Gas? World Energy Outlook 2011.* Paris, France: IEA.

IEA (International Energy Agency). (2012a). *Energy Statistics of OECD Countries 2012.* Paris, France: IEA.

IEA (International Energy Agency). (2012b). *Golden Rules for a Golden Age of Gas.* Paris, France: IEA.

IEA (International Energy Agency). (2013). *Energy Balances of OECD Countries (2013 Edition).* Paris, France: IEA.

Industry Canada. (2010). *Voluntary Codes Guide - What is a Voluntary Code?* Retrieved September 2013, from http://www.ic.gc.ca/eic/site/oca-bc. nsf/eng/ca00963.html.

Infante, L., Hopkins, J., Obenshain, K., & Fisher, E. (2012). *Emerging Natural Gas Issues: Concerns and Activities Surrounding Hydraulic Fracturing and the Development of Shale Gas.* Washington (DC): Edison Electric Institute.

Interorganizational Committee on Guidelines and Principles for Social Impact Assessment. (1994). *Guidelines and Principles for Social Impact Assessment.* Washington (DC): National Oceanic and Atmospheric Administration.

IPCC (Intergovernmental Panel on Climate Change). (2007). *Climate Change 2007 – The Physical Science Basis.* New York (NY): IPCC.

IPCC (Intergovernmental Panel on Climate Change). (2013). *Climate Change 2013: The Physical Science Basis.* New York (NY).

IPCC. Jackson, R. & Dusseault, M. (2013). Seepage Pathway Assessment for Natural Gas to Shallow Groundwater During Well Stimulation, Production, and After Abandonment. Paper presented at *GéoMontréal 2013*, Montréal (QC).

Jackson, R. B., Vengosh, A., Darrah, T. H., Warner, N. R., Down, A., Poreda, R. J., Karr, J. D. (2013a). Increased stray gas abundance in a subset of drinking water wells near Marcellus Shale gas extraction. *Proceedings of the National Academy of Sciences of the United States of America (PNAS),* *110*(8), 11250-11255. doi: 10.1073/pnas.1221635110.

Jackson, R.E., Gorody, A.W., Mayer, B., Roy, J.W., Ryan, M.C., & Van Stempvoort, D.R. (2013b). Groundwater protection and unconventional gas extraction: The critical need for field-based hydrogeological research. *Groundwater, 51*(4), 488-510. doi: 10.1111/gwat.12074.

Jacquet, J. (2009). *Energy Boomtowns and Natural Gas: Implications for Marcellus Shale Local Governments and Rural Communities. NERCRD* Rural Development Paper No. 43. University Park (PA): The Northeast Regional Center for Rural Development and Pennsylvania State University.

Jacquet, J. (2013). Risks to Communities from Shale Gas Development. Paper presented at *Workshop on Risks from Shale Gas Development,* Washington (DC).

Jewell, K. P. & Wilson, J. T. (2011). A new screening method for methane in soil gas using existing groundwater monitoring wells. *Ground Water Monitoring and Remediation, 31*(3), 82-94. doi: 10.1111/j1745-6592.2011.001345.x.

Jiang, M., Griffin, W. M., Hendrickson, C., Jaramillo, P., VanBriesen, J., & Venkatesh, A. (2011). Life cycle greenhouse gas emissions of Marcellus shale gas. *Environmental Research Letters, 6*(3), 034014.

Johnson, E. (2012). Water Issues Associated with Hydraulic Fracturing in Northeast British Columbia. Paper presented at *Unconventional Gas Technical Forum,* Victoria (BC).

Johnson, E. & Johnson, L. A. (2012). Hydraulic fracture water usage in northeast British Columbia: Locations, volumes and trends. Geoscience Reports 2012, B.C. *Ministry of Energy and Mines* 2012, 41-63.

Johnson, N. (2010). *Pensylvania Energy Impacts Assessment. Report 1: Marcellus Shale Natural Gas and Wind.* Harrisburg (PA): The Nature Conservancy.

Johnson, N. (2011). *Pennsylvania Energy Impacts Assessment.* Harrisburg (PA): The Nature Conservancy.

Jones, C. F. (2010). A landscape of energy abundance: Anthracite coal canals and the roots of american fossil fuel dependence, 1820–1860. *Environmental History, 15*, 449-484. doi: 10.1093/envhis/emq057.

Kahrilas, G., Blotevogal, J., Corrin, E. R., & Borch, T. (2013). *Fate of Hydraulic Fracturing Chemicals Under Down-Hole Conditions.* Paper presented at ACS National Meeting, Indianapolis (IN).

Kelly, W. R. & Mattisoff, G. (1985). The effects of a gas well blow out on groundwater chemistry. *Environmental Geology and Water Sciences, 7*(7), 205-213.

Kemball-Cook, S., Bar-Ilan, A., Grant, J., Parker, L., Jung, J., Santamaria, W., Yarwood, G. (2010). Ozone impacts of natural gas development in the Haynesville Shale. *Environmental Science and Technology, 44*(24), 9357-9363.

Keranen, K. M., Savage, H. M., Abers, G. A., & Cochran, E. S. (2013). Potentially induced earthquakes in Oklahoma, USA: Links between wastewater injection and the 2011 M_w 5.7 earthquake sequence. *Geology, 41*(6), 699-702.

King, G. E. (2012). *Hydraulic Fracturing 101: What Every Representative, Environmentalists, Regulator, Reporter, Investor, University Researcher, Neighbor and Engineer Should Know About Estimating Frac Risk and Improving Frac Performance in Unconventional Gas and Oil Wells.* Paper presented at SPE Hydraulic Fracturing Technology Conference, The Woodlands, Texas.

Korfmacher K. S., Jones W. A., Malone S. L., & Vinci L. F. (2013). Public health and high volume hydraulic fracturing. *New Solutions, 23*(1), 13-31.

Kresic, N. & Mikszewski, A. (2012). *Hydrogeological Conceptual Site Models: Data Analysis and Visualization.* Boca Raton (FL): CRC Press.

Kueper, B. W., Haase, C. S., & King, H. L. (1992). Leakage of dense, nonaqueous phase liquids from waste impoundments constructed in fractured rock and clay: Theory and case history. *Canadian Geotechnical Journal, 29*, 234-244.

Lackey, S. O., Myers, W. F., Christopherson, T. C., Gottula, J. J. (2009). *Nebraska Grout Task Force.* Lincoln (NE): University of Nebraska.

228 *References*

Lapierre, L. (2012). *The Path Forward. Part I: Public Meeting Summary.* Fredericton (NB): The Government of New Brunswick.

Lavoie, D., Chen, Z., Pinet, N., & Lyster, S. (2012). *A Review of November 24–25, 2011 Shale Gas Workshop. Resource Evaluation Methodology.* Calgary (AB): Geological Survey of Canada.

Lemieux, J. M. (2011). The potential impact of underground geological storage of carbon dioxide in deep saline aquifers on shallow groundwater resources. *Hydrogeology Journal, 19*, 457-778. doi: 10.1007/s10040-011-0715-4.

Liroff, R. (2012). *Extracting the Facts: An Investor Guide to Disclosing Risks from Hydraulic Fracturing Operations.* New York (NY): Investor Environmental Health Network and Interfaith Center on Corporate Responsibility.

Litovitz, A., Curtright, A., Abramzon, S., Burger, N., & Samaras, C. (2013). Estimation of regional air-quality damages from Marcellus Shale natural gas extraction in Pennsylvania. *Environmental Research Letters, 8*(1).

Logan, J., Heath, G., Paranhos, E., Boyd, W., & Carlson, K. (2012). *Natural Gas and the Transformation of the US.* Golden (CO): Joint Institute for Strategic Energy Analysis.

Lolon, E. P., Cipolla, C. L., & Weijers, L. (2009). *Evaluation Horizontal Well Placement and Hydraulic Fracture Spacing/Conductivity in the Bakken Formation, North Dakota.* Paper presented at Society of Petroleum Engineers Annual Technical Conference and Exhibition, New Orleans (LA).

Lowen Hydrogeology Consultants Ltd. (2011). *Aquifer Classification Mapping in the Peace River Region for the Montney Water Project.* Victoria (BC): Geoscience B.C.

Manga, M., Beresnev, I., Brodsky, E. E., Elkhoury, J. E., Elsworth, D., Ingebritsen, S. E., Wang, C. Y. (2012). Changes in permeability caused by transient stresses: Field observations, experiments and mechanisms. *Review of Geophysics, 50*, RG000382.

Manning, M. & Reisinger, A. (2011). Broader perspectives for comparing different greenhouse gases. *Philosophical Transactions of the Royal Society, 369*(1943), 1891-1905.

Martin, G. (2003). *Management of New Brunswick's Crown Forest*. Fredericton (NB): New Brunswick Department of Natural Resources.

Matthews, H. D., Gillett, N. P., Stott, P. A., & Zickfeld, K. (2009). The proportionality of global warming to cumulative carbon emissions. *Nature, 459*, 829-832.

Maule, A. L., Makey, C. M., Benson, E. B., Burrows, I. J., & Scammell, M. K. (2012). Disclosure of hydraulic fracturing fluid chemicals additives: Analysis of regulations. *New Solutions: A Journal of Environmental and Occupational Health Policy, 23*(1), 167-187.

McGarr, A., Simpson, D., & Seeber, L. (2002). 40 case histories of induced and triggered seismicity. *International Handbook of Earthquake and Engineering Seismology, 81*(A), 647-661.

McKenzie, L. M., Witter, R. Z., Newman, L. S., & Adgate, J. L. (2012). Human health risk assessment of air emissions from development of unconventional natural gas resources. *Science of the Total Environment, 424*(1), 79-87.

McLeod, R. (2011, July-August). $and dollars – mining frac sand in the River Valley, *Big River Magazine*.

Meyer, J. R., Parker, B. L., & Cherry, J. A. (2008). Detailed hydraulic head profiles as essential data for defining hydrogeologic units in layered fractured sedimentary rock. *Environmental Geology, 56*, 27-44.

Millennium Ecosystem Assessment. (2005). *Ecosystems and Human Well-Being*. Washington (DC): Island Press.

MIT (Massachussetts Institute of Technology). (2011). *The Future of Natural Gas*. Cambridge (MA): MIT.

Moffet, J., Bregha, F., & Middelkoop, M. J. (2004). Responsible Care: A Case Study of a Voluntary Environmental Initiative. In K. Webb, *Voluntary Codes*. Ottawa (ON): Carleton Research Unit for Innovation, Science and Environment.

Muehlenbachs, K. (2012a). Using Stable Isotope Geochemistry to Fingerprint Fugitive Gases from Hydraulically Fractured Wells. Paper presented at *Hydraulic Fracture Stimulation: Science, Society & Environment, Canadian Society of Petroleum Geologists' Gussow Conference*, Banff (AB).

Muehlenbachs, K. (2012b). Identifying the Sources of Fugitive Methane Associated with Shale Gas Development, Updated with New Data, Jan 2012. Paper presented *at Resource for the Future. Managing the Risks of Shale gas: Identifying a Pathway Toward Responsible Development,* Washington (DC).

Mueller, D. & Eid, R. (2006). Characterizing Early-Stage Physical Properties, Mechanical Behavior of Cement Designs. Paper presented at *Drilling Conference,* Miami Beach (FL).

Mufson, S. (2012, February 1). Cheap Natural Gas Jumbles Energy Markets, Stirs Fears it Could Inhibit Renewables, *Washington Post.*

Myers, T. (2012). Potential contaminant pathways from hydraulically fractured shale to aquifers. *Groundwater, 50*(6), 872-882.

NETL (National Energy Technology Laboratory). (2010). *Improving Thermal Efficiency of Power Plants. Technical Workshop Report.* Baltimore (MD): U.S. Department of Energy.

Neuzil, C. E. (1986). Groundwater flow in low-permeability environments. *Water Resources Research, 22*(8), 1163-1195.

Neuzil, C. E. (1994). How permeable are clays and shales? *Water Resources Research, 30*(2), 145-150.

Nielsen, A. (2012). *We Gambled Everything: The Life and Times of an Oilman.* Edmonton (AB): The University of Alberta Press.

NIOSH (National Institute for Occupational Safety and Health). (2002). *Health Effects of Occupational Exposure to Respirable Crystalline Silica.* Cincinnati (OH): Department of Health and Human Services, Centers for Disease Control and Prevention, NIOSH.

Northrup, J. M. & Wittemyer, G. (2013a). Characterising the impacts of emerging energy development on wildlife, with an eye towards mitigation. *Ecology Letters, 16(1),* 112-125. doi: 10.1111/ele.12009.

NYSDEC (New York State Department of Environmental Conservation). (2011). *Revised Draft Supplemental Generic Environmental Impact Statement on the Oil, Gas and Solution Mining Regulatory Program.* Albany (NY): NYSDEC.

O'Sullivan, F. & Paltsev, S. (2012). Shale gas production: Potential versus actual greenhouse gas emissions. *Environmental Research Letters, 7*(4). doi: doi:10.1088/1748-9326/7/4/044030.

OCMOH (Office of the Chief Medical Officer of Health). (2012). *Chief Medical Officer of Health's Recommendations Concerning Shale Gas Development in New Brunswick.* Fredericton (NB): New Brunswick Department of Health.

Ordonez, I. (2012). *Oil Companies Drawn to 'Frac Sand'.* Retrieved July 2013, from http://www.rigzone.com/news/oil_gas/a/117839/ Oil_ Companies_Drawn_To_Frac_Sand.

Oreskes, N. & Belitz, K. (2001). Philosophical Issues in Model Assessment. In M. G. Anderson & P. D. Bates, *Model Validation: Perspectives in Hydrological Science.* London, United Kingdom: John Wiley and Sons Ltd.

Oreskes, N. (2011). Working with Uncertainty: 'Unitisation and Renegotiation' as a Model for Science and Environmental Policy. In J. Lentsch & P. Weingart, *The Politics of Science Advice: Institutional Design for Quality Assurance.* Cambridge, United Kingdom: Cambridge University Press.

Osborn, S. G. & McIntosh, J. C. (2010). Chemical and isotopic tracers of the contribution of microbial gas in Devonian organic-rich shales and reservoir sandstones, northern Appalachian Basin. *Applied Geochemistry, 25*, 456-471.

Osborn, S. G., Vengosh, A., Warner, N. R., & Jackson, R. B. (2011). Methane contamination of drinking water accompanying gas-well drilling and hydraulic fracturing. *Proceedings National Academy of Science, 108*, 8172-8176. doi: 10.1073/pnas.1100682108.

OSHA (Occupational Safety and Health Administration) & NIOSH (National Institute for Occupational Safety and Health). (2012). *Hazard Alert. Worker Exposure to Silica During Hydraulic Fracturing.* OSHA & NIOSH.

Patton, F. D. & Smith, H. R. (1988). Designs Considerations and the Quality of Data from Multiple Level Ground-Water Monitoring Wells. In A. G. Collins & A. I. Johnson, *Ground-Water Contamination: Field Methods.*

232 *References*

ASTM STP 963. Philadelphia (PA): American Society for Testing and Materials.

PennEnergy Editorial Staff. (2013, January 11). Halliburton using clean-burning natural gas to power fracking fleet, *PennEnergy*.

Pennsylvania Office of the Governor. (2012). *Governor Corbett Signs Historic Marcellus Shale Law.* Retrieved May 2013, from http://www.pikepa.org/Planning/Marcellus/Governor%20Corbett%20S igns%20Historic%20%20Marcellus%20Shale%20Law.pdf.

Perry, S. L. (2013). Using ethnography to monitor the community health implications of onshore unconventional oil and gas developments: Examples from Pennsylvania's Marcellus Shale. *New Solutions: A Journal of Environmental and Occupational Health Policy, 23*(1), 33-53.

Petron, G., Frost, G., Miller, B. T., Hirsch, A. I., Montzka, S. A., Karion, A., Hall, B. (2012). Hydrocarbon emissions characterization in the Colorado Front Range – a pilot study. *Journal of Geophysical Research: Atmospheres 117*(D4), D04 304. doi: 10.1029/2011JD016360.

Pochon, A., Tripet, J.-P., Kozel, R. M., B., Sinreich, M., & Zwahlen, F. (2008). Groundwater protection in fractured media: A vulnerability-based approach for delineating protection zones in Switzerland. *Hydrogeology Journal, 16*(7), 1267-1281. doi: 10.1007/s10040-008-0323-0.

Porter, M. E. & Kramer, M. R. (2006). Strategy and society: The link between competitive advantage and corporate social responsibility. *Harvard Business Review, 2006*, 78-92.

Precht, P. & Dempster, D. (2012). *Jurisdictional Review of Hydraulic Fracturing Regulation.* Halifax (NS): Nova Scotia Department of Energy and Nova Scotia Environment.

Province of British Columbia. (2012). B.*C. Regulation 282/2010. Oil and Gas Activities Act. Drilling and Production Regulation.* Retrieved January 2014, from http://www.bclaws.ca/EPLibraries/bclaws_new/ document/ID/freeside/282_2010.

Pruess, K., Oldenburg, C., & Moridis, G. (1999). *TOUGH2 User's Guide, Version 2*. Berkeley (CA): Earth Sciences Division, Lawrence Berkeley National Laboratory, University of California.

Pruess, K. (2005). *ECO2N: A TOUGH2 Fluid Property Module for Mixtures of Water, NaCl and CO_2*. Berkeley (CA): Earth Sciences Division, Lawrence Berkeley National Laboratory, University of California.

PwC (PricewaterhouseCoopers LLP). (2011). *Shale Gas. A Renaissance in U.S. Manufacturing?* PwC.

Raleigh, C. B., Healy, J. H., & Bredehoeft, J. D. (1976). An Experiment in earthquake control at rangely, Colorado. *Science, 191*(4233), 1230-1237.

Raven, K. G., Novakowksi, K. S., Yager, R. M., & Heystee, R. J. (1992). Supernormal fluid pressures in sedimentary rocks of southern Ontario – western New York State. *Canadian Geotechnical Journal, 29*, 80-93.

République Française. (2013). L'exploitation et l'exploration des gaz de schiste définitivement interdites. [Exploitation and exploration of shale gas definitely prohibited.] Retrieved January 2014, from http://www.gouvernement.fr/gouvernement/en-direct-des-ministeres/fracturation-hydraulique-l-exploitation-et-l-exploration-des-g.

Resnikoff, M. (2012). *Radon in Natural Gas from Marcellus Shale*. New York (NY): Submission to the New York State Department of Environmental Conservation.

Rich, A. L. & Crosby, E. C. (2013). Analysis of reserve pit sludge from unconventional natural gas hydraulic fracturing and drilling operations for the presence of technologically enhanced naturally occurring radioactive material (TENORM). *New Solutions: A Journal of Environmental and Occupational Health Policy, 23*(1), 117-135. doi: 10.2190/NS.23.1.h.

Richardson, B. (2013). Natural Gas Is Bridge from Oil to Renewables. *Bloomberg TV*. Retrieved July 2013, from http://www.youtube.com/watch?v=-Br9BPVcrFo.

Rivard, C., Molson, J., Soeder, D. J., Johnson, E. G., Grasby, S., Wang, B., & Rivera, A. (2012). *A Review of the November 24–25, 2011 Shale Gas*

234 *References*

Workshop, Calgary, Alberta – 2. Groundwater Resources. Open File 7096. Natural Resources Canada.

Roy, J. W. & Ryan, M. C. (2013). Effects of unconventional gas development on groundwater: A call for total dissolved gas pressure field measurements. *Groundwater, 51*(4), 480-482.

Royte, E. (2012, November 28). Fracking Our Food Supply, *The Nation*.

RSI (Risk Sciences International). (2012). *An Assessment of the Lung Cancer Risk Associated with the Presence of Radon in Natural Gas Used for Cooking in Homes in New York State.* Ottawa (ON): RSI.

Ryder Scott Company Petroleum Consultants. (2008). *Appendix A: Resource Potential Horton Bluff Formation, Windsor Basin.* Calgary (AB): Ryder Scott Company Petroleum Consultants.

Saiers, J. E. & Barth, E. (2012). Potential contaminant pathways from hydraulically fractured shale aquifers. *Ground Water, 50*(6), 826-828.

Schrag, D. P. (2012). Is shale gas good for climate change? *Daedalus, 41*(2), 72-80. SEAB (Secretary of Energy Advisory Board). (2011a). *Shale Gas Production Subcommittee Second Ninety Day Report.* Washington (DC): U.S. Department of Energy.

SEAB (Secretary of Energy Advisory Board). (2011b). *Shale Gas Production Subcommittee 90-Day Report.* Washington (DC): U.S. Department of Energy.

Séjourné, S., Lefebvre, R., Lavoie, D., & Malet, X. (2013). *Geologic and Hydrogeoloic Understanding of Caprock Strata Overlying the Utica Shale in the St. Lawrence Lowlands (Quebec, Canada).* Paper presented at GéoMontréal 2013, Montreal (QC).

Sethi, S. P. (1979). A conceptual framework for environmental analysis of social issues and evaluation of business response pattern. *Academy of Management Review, 4*(1), 63-74.

Shindler, B. & Toman, E. (2003). Fuel reduction strategies in forest communities: A longitudinal analysis. *Journal of Forestry, 101*(6), 8-14.

Shindler, B. A., Brunson, M. W., & Stankey, G. H. (2002). *Social Acceptability of Forest Conditions and Management Practices: A*

Problem Analysis. Portland (OR): U.S. Department of Agriculture Pacific Northwest Research Station.

Siemens. (2013). *Siemens to Build a Turnkey Combined Cycle Power Plant in the Philippines.* Retrieved January 2014, from http://www.siemens.com/press/en/pressrelease/?press=/en/pressrelease/2013/energy/power-generation/ep201312015.htm.

Sierra Research Inc. (2011). *Screening Health Risk Assessment Sublette County, Wyoming.* Sacramento (CA): Sublette County Commissioners, Wyoming Department of Environmental Quality and Wyoming Department of Health.

Slonecker, E. T., Milheim, L. E., Roig-Silva, C. M., Malizia, A. R., Marr, D. A., & Fisher, G. B. (2012). *Landscape Consequences of Natural Gas Extraction in Bradford and Washington Counties, Pennsylvania, 2004–2010.* Reston (VA): U.S. Geological Survey.

Soeder, D. J. (1988). Porosity and permeability of eastern devonian gas shale. *Society of Petroleum Engineers*, 116-124.

Staaf, E. (2012). *Risky Business: An Analysis of Marcellus Shale Gas Drilling Violations in Pennsylvania 2008–2011.* Pittsburgh (PA): PennEnvironment Research and Policy Center.

Stark, M., Allingham, R., Calder, J., Lennartz-Walker, T., Wai, K., Thompson, P., & Zhao, S. (2012). *Water and Shale Gas Development.* Accenture Consulting.

Subra, W. (2009). *Results of Health Survey of Current and Former DISH/Clark, Texas Residents.* New Iberia (LA): Earthworks.

The Royal Society and Royal Academy of Engineering. (2012). *Shale Gas Extraction in the U.K.: A Review of Hydraulic Fracturing.* London, United Kingdom: The Royal Society and the Royal Academy of Engineering.

Tilley, B. & Muehlenbachs, K. (2011). Fingerprinting of Gas Contaminating Groundwater and Soil in a Petroliferous Region, Alberta, Canada. Paper presented at *International Network of Environmental Forensics Conference*, Cambridge, United Kingdom.

236 *References*

Tilley, B. & Muehlenbachs, K. (2013). Isotope reversals and universal stages and trends of gas maturation in sealed, self-contained petroleum systems. *Chemical Geology, 339,* 194-204.

Tollefson, J. (2013a). Nature News: Methane leaks erode green credentials of natural gas. *Nature, 493*(7430), 12.

Tollefson, J. (2013b). Methane leaks erode green credentials of natural gas. *Nature, 493*(12).

Triangle Petroleum. (2009). *Corporate Profile 2009.* Calgary (AB): Triangle Petroleum.

U.K. Parliament. (1996). *Offshore Installations and Wells (Design and Construction, etc) Regulations 1996.* London, United Kingdom.

Unger, A. J. A., Sudicky, E. A., & Forsyth, P. A. (1995). Mechanisms controlling vacuum extraction coupled with air sparging for remediation of heterogeneous formations contaminated by dense nonaqueous phase liquids. *Water Resources Research, 31*(8), 1913-1915. doi: 10.1029/95WR00172.

URS. (2012). *Revised Attachment 3: Gas Well Completion Emissions Data. Submitted as part of ANGA & AXPC. (2012). ANGA Comments to EPA on New Source Performance Standards for Hazardous Air Pollutants Review America's Natural Gas Alliance.* Washington (DC): ANGA and AXPC.

Van Stempvoort, D., Maathuis, H., aworski, E., Mayer, B., & Rich, K. (2005). Oxidation of fugitive methane in ground water linked to bacterial sulfate reduction. *Groundwater, 43*(2), 187-199. doi: 10.1111/j.1745-6584.2005.0005.x.

Van Stempvoort, D. & Roy, J. W. (2011). *Potential Impacts of Natural Gas Production on Groundwater Quality in Canada and Related Research Needs.* Ottawa (ON): Environment Canada.

Vaughn, S. (2012). *Report of the Commissioner of the Environment and Sustainable Development.* Ottawa (ON): Office of the Auditor General of Canada.

Vengosh, A., Warner, N., Jackson, R., & Darrah, T. (2013). The effects of shale gas exploration and hydraulic fracturing on the quality of water resources in the United States. *Procedia Earth and Planetary Science, 7*(2), 863-866.

Venkatesh, A., Jaramillo, P., Griffin, W. M., & Matthews, H. S. (2011). Uncertainty in life cycle greenhouse gas emissions from United States natural gas end-uses and its effects on policy. *Environmental Science and Technology, 45*(19), 8182-8189. doi: 10.1021/es200930h.

Vidic, R., Brantley, S. L., Vandenbossche, J. M., Yoxtheimer, D., & Abad, J. D. (2013). Impact of shale gas development on regional water quality. *Science, 340*(6134), 1235009.

Walsh, W., Adams, C., Kerr, B., & Korol, J. (2006). *Regional "Shale Gas" Potential of the Triassic Doig and Montney Formations, Northeastern British Columbia.* Victoria (BC): British Columbia Ministry of Energy, Mines and Petroleum Resources.

Wang, B. (2013a). A Hypothetical Geomechanics Model for Qualitative Assessment of Environmental Impact of Shale Gas Fracking – Part I: Potential Effect on Groundwater. Paper presented at *GéoMontréal 2013*, Montréal (QC).

Wang, B. (2013b). A Hypothetical Geomechanics Model for Qualitative Assessment of Environmental Impact of Shale Gas Fracking – Part II: Implication for Induced Seismicity. Paper presented at *GéoMontréal 2013*, Montréal (QC).

Warner, N. R., Jackson, R. B., Darrah, T. H., Osborn, S. G., Down, A., Zhao, K., Vengosh, A. (2012). Geochemical evidence for possible natural migration of Marcellus Formation brine to shallow aquifers in Pennsylvania. *Proceeding of the National Academy of Sciences, 109*(30), 11961-11966.

Warner, N. R., Christie, C. A., Jackson, R. B., & Vengosh, A. (2013). Impacts of shale gas wastewater disposal on water quality in western Pennsylvania. *Environmental Science and Technology, 47*(20), 11849-11857.

Watson, M. (2003). Environmental Impact Assessment and European Community Law. Paper presented at *XIV International Conference – Danube River of Cooperation*, Belgrade, Serbia.

238 *References*

Watson, T. L. (2004). Surface Casting Vent Flow Repair – A Process. Paper presented at *5th Canadian International Petroleum Conference*, Calgary (AB).

Watson, T. L. & Bachu, S. (2009). Evaluation of the potential for gas and CO_2 leakage along wellbores. *SPE Drilling & Completion Journal 24*(1), 115-126.

Webb, K. (2004). Understanding Voluntary Codes. In K. Webb, *Voluntary Codes: Private Governance, the Public Interest, and Innovation*. Ottawa (ON): Carleton Research Unit for Innovation, Science and Environment.

Weber, C. L. & Clavin, C. (2012). Life cycle carbon footprint of shale gas: Review of evidence and implications. *Environmental Science and Technology, 46*(11), 5688-5695.

Whitaker, M., Heath, G. A., O'Donoughue, P., & Vorum, M. (2012). Life Cycle Greenhouse Gas Emissions of Coal-Fired Electricity Generation. *Journal of Industrial Ecology, 16*(51), S53-S72.

WHO (World Health Organization). (1946). *Constitution of the World Health Organization*. Vol. 36. Copenhagen, Denmark: WHO.

WHO (World Health Organization). (2009). *Milestones in Health Promotion. Statements from Global Conferences*. Geneva, Switzerland: WHO.

Wigley, T. M. L. (2011). Coal to gas: The influence of methane leakage. *Climatic Change, 108*, 601-608. doi: http://dx.doi.org/10.1007/s10584-011-0217-3.

Williams, H. F. L., Havens, D. L., Banks, K. E., & Wachal, D. J. (2008). Field-based monitoring of sediment runoff from natural gas well sites in Denton County, Texas, USA. *Environmental Geology, 55*, 1463-1471.

Wilson, J. M. & VanBriesen, J. (2012). Oil and gas produced water management and surface drinking water sources in Pennsylvania. *Environmental Practice, 14*(4), 288-300. doi:10.10170S1466 046612000427.

Wilson, S., Subra, W., & Sumi, L. (2013). *Reckless Endangerment While Fracking the Eagle Ford*. Washington (DC): Earthworks.

Witter, R., Stinson, K., Sackett, H., Putter, S., Kinney, G., Teitelbaum, D., & Newman, L. (2008). *Potential Exposure-Related Human Health*

References

Effects of Oil and Gas Development: A Literature Review (2003–2008). Denver (CO): Colorado School of Public Health.

Witter, R., McKenzie, L., Towle, M., Stinson, K., Scott, K., Newman, L., & Adgate, J. (2010). *Health Impact Assessment for Battlement Mesa, Garfield County Colorado*. Aurora (CO): Colorado School of Public Health, University of Colorado.

Wolf Eagle Environmental. (2009). *Town of DISH, Texas. Ambient Air Monitoring Analysis*. Flower Mound (TX): Wolf Eagle Environmental.

Worster, D. (1992). *Under Western Skies: Nature and History in the American West*. New York (NY): Oxford University Press.

Wyoming Department of Environmental Quality. (2008). *Compliance Monitoring and Siting Requirements for Unlined Impoundments Receiving Coalbed Methane Produced Water*. Laramie (WY): Wyoming Department of Environmental Quality.

Xu, T., Spycher, N., Sonnenthal, E., Zheng, L., & Pruess, K. (2012). *TOUGHREACT User's Guide: A Simulation Program for Non-isothermal Multiphase Reactive Transport in Variably Saturated Geologic Media, Version 2.0*. Berkeley (CA): Earth Sciences Division, Lawrence Berkeley National Laboratory.

Young, O. R., King, L. A., & Schroeder, H. (2008). *Institutions and Environmental Change*. Cambridge (MA): MIT Press.

Yoxtheimer, D. (2012). *Addressing the Environmental Risks from Shale Gas Development*. Washington (DC): Worldwatch Institute.

Zeirman, R. (2013, June 21). Why Such Hysteria Over Fracking? *Los Angeles Times*. Zhou, Q., Birkholzer, J. T., Mehnert, E., Lin, Y.-F., & Zhang, K. (2010). Modeling basinand plume-scale processes of CO_2 storage for full-scale deployment. *Ground Water, 48*(4), 494-514.

Zoback, M. (2010). *Reservoir Geomechanics*. Cambridge, United Kingdom: Cambridge University Press.

Zoback, M., Kitasei, S., & Copithorne, B. (2010). *Addressing the Environmental Risks from Shale Gas Development*. Washington (DC): WorldWatch Institute.

Zoback, M. (2012). Managing the seismic risk posed by wastewater disposal. *Earth Magazine, 57*(4), 38-44.

ABOUT THE AUTHOR

Afsoon Moatari-Kazerouni
Environmental Engineering and Management Research Group
Ton Duc Thang University
Ho Chi Minh City, Vietnam
and Faculty of Environment and Natural Resources
Department of Water Science and Technology
Ton Duc Thang University
Ho Chi Minh City, Vietnam
Email: afsoon.moatari-kazerouni@tdt.edu.vn

INDEX

A

abandoned wells, ix, xiii, 6, 12, 35, 60, 69, 80, 82, 93, 200

abandonment, 8, 90, 93, 111, 146, 174, 225

Aboriginal issues, 164

abstraction, 73, 81

acceptability, xii, 10, 13, 190, 234

access roads, 21, 102

acoustic signals, 31

adaptive management, 109, 166, 196

additive, 40, 41, 127, 181, 205, 206, 215

aesthetic indicators, 155

aesthetical, 146

aggravated asthma, 121

agricultural activities, 51, 140

agricultural land, 8, 20, 122, 188

agriculture, xi, 4, 52, 107, 235

Air Contaminants, xii

air emissions, 3, 20, 40, 85, 97, 98, 115, 126, 189, 191, 220, 229

air quality, xi, xii, 21, 73, 98, 138, 170

anaerobic bacteria, 73

ancillary facilities, 3, 4

annular gas flow, 149

annulus, 29, 30, 31, 44, 58, 69, 93, 145, 149, 152, 153, 205

anthropogenic, viii, x, 10, 50, 68, 85, 197

anthropogenic climate change, viii, x, 85, 197

aquatic habitat, 107

aquifer invasion, 149

aquifer vulnerability, 130, 156, 157, 158

aquifer(s), ix, xv, 12, 17, 30, 38, 46, 51, 56, 70, 71, 74, 82, 130, 146, 149, 154, 155, 156, 157, 158, 163, 198, 200, 205, 209, 216, 228, 230, 234

aquitard integrity, 157

aquitard(s), 157, 205

aromatics, x, 14, 201

arsenic, 73, 78, 119

assimilation capacity, ix, 71, 72, 74, 82, 156, 200

atmosphere, xiv, xv, 15, 24, 29, 70, 73, 82, 88, 89, 133, 134, 135

atmospheric aerosols, 96

atmospheric sulphate aerosols, 86

auger rig, 29

244 *Index*

B

bacteria, 17, 40, 73, 155, 205
barium, 74, 78, 117
baseline monitoring, 75, 82, 127, 146, 155, 166
Basin, 120, 122, 123, 125, 205, 213, 221, 222, 224, 231, 234
bedding fabric, 35
bedding planes, 61, 62
bedrock, 61, 65, 74, 155, 205
behavioural characteristics, 53
behavioural impairment, 121
Bentonite-based, 32
bentonite-water slurry, 29
benzene, x, xii, 14, 52, 98, 120, 121, 201
berry-picking, 21
bioaccumulation, 79
biocide, 40, 205
biodegradation, 72, 148
biofuels, 6
biogenic, 17, 18, 58, 153, 156, 205
biogenic methane, 17, 18, 58, 205
biogeochemical reactions, 73, 149
biomass productivity, 107
blow-out preventer (BOP), 29, 206
booming population, 123
boreal forest, 8
borehole logging, 143
borehole tests, 63, 65
bottom-hole assembly fabrication, 32
brackish, 50, 56, 58, 77, 78
brackish water, 50, 77
brine(s), 30, 56, 57, 58, 61, 62, 66, 119, 127, 149, 151, 206, 213, 216, 221, 237
bromine, 74
BTEX, 97, 98, 119, 121
BTEX compounds, 119
bulk permeability, 63, 65
buoyancy, 58, 73, 82, 149
buoyant, 61, 62, 68, 71, 93

butane, 17, 26, 32, 36, 58

C

calcium, 74, 206
cancer, 119, 120, 126, 234
capacity concept, 70
caprocks, 35, 82
carbon dioxide, x, 6, 8, 17, 19, 32, 36, 38, 85, 86, 87, 88, 89, 90, 91, 94, 96, 98, 110, 135, 151, 180, 228
carbon dioxide injection, 110, 151
carbon-free, 1
carbon-intensive, 4
carbon-neutral, 6
casing, 14, 24, 28, 29, 30, 31, 35, 43, 44, 45, 46, 55, 69, 72, 74, 93, 99, 134, 135, 136, 137, 138, 153, 154, 155, 205, 206, 207, 208, 209, 210, 214, 216
casing leakage, 55
Casing Shoe, 206
casing string(s), 28, 30, 31, 206, 207, 208
CCS site, 141
cement, viii, xiv, 14, 27, 29, 30, 31, 43, 44, 45, 46, 55, 58, 69, 72, 93, 134, 135, 144, 149, 153, 154, 155, 167, 174, 199, 206, 207, 210, 217, 218, 230
cement bond log(s), 31, 45, 72, 206
cement job, 206
cement seal(s), viii, xiv, 14, 43, 44, 46, 55, 58, 72, 134, 149, 167, 199
cement-seal annulus, 144
centralized treatment plants, 38
centralizer, 206
chemical, xiii, xiv, 6, 8, 10, 11, 21, 23, 25, 36, 38, 39, 44, 47, 53, 56, 60, 66, 78, 79, 81, 97, 98, 119, 122, 140, 148, 150, 154, 156, 157, 172, 177, 179, 182, 191, 205, 206, 216, 223, 231, 236
Chemical Abstracts Service, 206
chemical and, 56, 60, 122, 231

Index

245

chemical composition, xiv, 8, 66, 78, 79, 119, 148, 154

chemical ingredient, 205, 206

chemical risks, 10

chemicals, viii, ix, x, xiv, 3, 4, 14, 19, 24, 27, 36, 38, 39, 40, 50, 51, 52, 53, 54, 55, 66, 67, 71, 72, 81, 82, 97, 115, 118, 119, 124, 127, 131, 138, 139, 140, 145, 148, 186, 199, 200, 216, 227, 229

chlorinated solvents, 63

circumstances, xii, 44, 68, 71, 130, 199

clay, 17, 18, 38, 40, 41, 62, 78, 222, 227

climate change, viii, x, 1, 5, 6, 20, 85, 86, 87, 88, 89, 92, 96, 98, 173, 198, 234

CNG engines, 189

coal, x, 1, 4, 6, 19, 74, 86, 89, 90, 92, 93, 94, 95, 96, 118, 202, 214, 223, 227, 238

combination, xii, 3, 24, 37, 71, 82, 202, 205

combined-cycle, 94, 97

combustion technology, 96

commercial, 18, 19, 23, 29, 36, 44, 50, 56, 90, 93, 112, 210

communication, 34, 69, 70, 120, 188, 195, 206, 217

community disruption, 14, 115, 192

community impacts, xi, 122

completion, 18, 23, 28, 31, 32, 37, 45, 50, 90, 91, 96, 102, 103, 108, 124, 138, 175, 177, 185, 189, 206, 208, 222, 236, 238

compliance monitoring, 144, 145

compressor, xi, 92, 101, 104, 118

compressor stations, xi, 92, 104

compressors, 98

concentration, xiv, 42, 127, 137, 186, 209, 223

conceptual models, 142, 151, 158

condensates, 53, 66

conductor pipe, 29, 31

congestion, 124

consumption, 76, 86, 93, 99

contaminant dispersion, 98

contaminant plume, 72

contaminants, viii, 47, 49, 50, 51, 52, 63, 69, 71, 81, 82, 120, 126, 127, 148, 153, 156, 157, 170, 197, 200

contamination, viii, xvi, 12, 13, 19, 20, 21, 29, 32, 43, 51, 52, 55, 56, 58, 59, 60, 61, 63, 67, 68, 69, 70, 71, 73, 75, 78, 80, 82, 85, 118, 120, 129, 137, 139, 140, 142, 143, 145, 152, 153, 154, 155, 156, 157, 160, 174, 198, 199, 201, 221, 231

contamination pathways, 59, 67, 80, 85

contaminative potential, 57

conventional, viii, ix, xii, xv, 2, 5, 6, 14, 15, 17, 18, 19, 22, 24, 26, 27, 35, 44, 45, 46, 51, 67, 69, 76, 77, 78, 90, 91, 92, 93, 94, 97, 101, 102, 111, 115, 118, 124, 137, 142, 146, 147, 186, 199, 210

conventional gas, ix, xii, 2, 5, 6, 14, 17, 19, 24, 26, 44, 76, 77, 91, 94, 97, 101, 102, 124, 186

conventional monitoring, 137, 147

conventional oil and gas, viii, ix, xv, 15, 27, 51, 67, 69, 101, 111, 115, 200

core, 64, 68, 143

Corrosion Inhibitor, 206

cottaging, 104

critical wildlife habitat, xiii, 108

cultural contexts, 5

cultural ramifications, 14

cumulative, ix, xi, xiii, xv, xvi, xvii, 4, 5, 8, 12, 13, 21, 45, 47, 55, 88, 101, 109, 120, 122, 123, 126, 127, 132, 156, 161, 166, 168, 170, 171, 186, 188, 200, 203, 212, 229

cumulative cancer risk, 120

D

data acquisition services, 32

data management, 32

Deep Zone, 32, 50, 52, 56, 57, 69, 70, 156

Deep-Well Disposal, 80

246 *Index*

deforestation, xi
deployment, xiii, 5, 111, 239
diesel aggregates, 98
diesel engines, 98, 189
diffusion, 71, 147
disposal, vii, x, 14, 25, 37, 49, 53, 75, 79, 80, 112, 113, 164, 181, 197, 206, 207, 215, 222, 237, 239
Disposal Well, 206
dissolved, 17, 18, 37, 40, 50, 60, 61, 70, 71, 73, 78, 135, 150, 154, 157, 207, 209, 210, 234
dissolved oxygen, 70, 71
distribution, x, 1, 10, 21, 50, 110, 137, 217, 219
divalent cations, 37
dolostone, 56
domestic, 9, 20, 60, 72, 73, 74, 82, 130, 140, 144, 153, 154, 155, 157, 159, 166, 176, 206
domestic water well(s), 130, 159
domestic well(s), 60, 72, 73, 74, 82, 140, 144, 153, 154, 155, 157, 159, 166, 176
downstream, x, 50, 79
drill cuttings, 31, 32, 119, 185
drill cuttings disposal, 32
drilling, vii, xi, xiv, xv, xvi, 8, 14, 22, 23, 26, 27, 28, 29, 30, 31, 38, 43, 44, 45, 46, 50, 51, 53, 54, 56, 58, 61, 63, 69, 70, 73, 77, 90, 97, 102, 103, 104, 118, 121, 124, 125, 137, 141, 142, 147, 156, 171, 172, 177, 181, 184, 185, 189, 193, 197, 198, 199, 206, 207, 210, 213, 214, 223, 230, 232, 233, 235, 238
drilling fluids, 32
drilling rigs, 97
DuMU^X, 150
dust, 20, 21, 105, 123, 124, 183, 188, 192

E

earthquakes, viii, xii, 19, 65, 109, 112, 113, 197, 201, 224, 227
ECLIPSE, 150
ecological, xv, 79, 106, 132, 216
ecological system, xv, 79, 132
economic, xi, 1, 4, 5, 6, 7, 11, 14, 22, 37, 63, 76, 77, 78, 106, 115, 116, 123, 124, 131, 132, 175, 182, 188, 198, 201, 202, 204
economic limitations, 37
economical, 23, 93, 189
economically, vii, 3, 18, 19, 202, 205, 208
ecosystem, 76, 101, 105, 106, 107, 122, 131, 204, 229
ecosystem integrity, 105
ecosystem services, 76, 106
ecosystems, xi, 76, 102, 106, 107, 204, 217, 229
electrical lines, 104
electricity, x, 86, 87, 89, 90, 94, 95, 96, 97, 118, 202, 225, 238
emergency plan, 131
emergency preparedness, 123, 202
emission(s), vii, x, xii, xiv, xv, 6, 12, 21, 47, 85, 86, 87, 88, 89, 90, 91, 92, 93, 94, 95, 96, 97, 98, 101, 107, 120, 124, 131, 134, 135, 138, 174, 183, 197, 198, 202, 211, 212, 214, 219, 223, 224, 226, 229, 231, 232, 236, 237
endocrine disruptors, 126
energized fluids, 36
energy, vii, x, 1, 3, 5, 6, 7, 17, 19, 22, 26, 40, 61, 67, 77, 86, 87, 89, 94, 95, 97, 101, 104, 109, 112, 122, 158, 173, 184, 190, 197, 198, 202, 203, 204, 211, 213, 215, 216, 218, 222, 224, 225, 226, 227, 228, 230, 232, 234, 235, 237
engineering disciplines, 63

Index

environmental impacts, viii, xiii, xv, xvii, 2, 3, 7, 8, 9, 10, 11, 12, 13, 16, 26, 55, 58, 75, 99, 101, 105, 129, 140, 141, 144, 145, 150, 160, 165, 166, 167, 169, 170, 171, 172, 173, 189, 197, 198, 199, 203, 204, 217

environmental risk(s), xiv, xvi, 7, 10, 13, 19, 24, 67, 79, 85, 89, 158, 171, 172, 176, 187

epidemiological studies, 127

Equity-focused health impact assessments (EFHIAs), 132

erosion, 107, 139

ethane, 17, 26, 58, 144

F

fate, viii, ix, 47, 66, 148, 161, 227

fault(s), ix, 8, 34, 46, 56, 60, 61, 62, 65, 67, 82, 109, 110, 111, 113, 138, 151, 158, 159, 200, 201, 221

feasible, vii, x, 37, 53, 148, 201, 202, 208

filing requirements, 175, 184

fingerprinting, 60, 153, 235

fishing, 104, 107, 123, 188

flammable, 36

flares, 97

flow barriers, 62

flow paths, 62, 68

flowback, ix, x, xiv, 3, 14, 25, 35, 36, 37, 38, 49, 50, 51, 53, 61, 67, 74, 77, 78, 79, 91, 97, 115, 117, 119, 120, 121, 145, 148, 185, 199, 200, 207, 208

flowback water, ix, x, xiv, 3, 14, 25, 37, 38, 49, 50, 51, 53, 61, 77, 78, 79, 91, 97, 116, 117, 119, 121, 145, 148, 199, 200

fluid evaporation, 98

fluid injection, 109, 113, 139

fluid migration, xi, 63, 164

fluid movement, 62, 149

fluids, viii, ix, 3, 7, 12, 15, 24, 25, 28, 29, 31, 35, 36, 37, 38, 40, 42, 44, 53, 54, 55, 56, 59, 61, 66, 70, 72, 79, 80, 81, 82, 91, 99, 110, 112, 113, 122, 145, 148, 151, 156, 180, 185, 191, 198, 206, 207, 208, 211, 216

forest plants, 106

formation, 4, 14, 18, 24, 27, 30, 34, 36, 37, 38, 41, 53, 62, 63, 66, 68, 72, 77, 78, 91, 93, 105, 110, 111, 135, 151, 152, 185, 206, 207, 208, 209, 220, 228, 234, 237

formation fluids, 66, 185, 206

fossil, vii, x, 6, 19, 58, 85, 86, 87, 202, 227

fossil fuel, x, 6, 19, 85, 86, 87, 202, 227

frac sand, 42, 229

fraction, x, 92

fracture networks, 62, 63, 65, 150

fracture stimulation, 26, 35, 36, 70

fracture zones, 67

fractured bedrock, 63

fractures, 15, 18, 24, 27, 28, 32, 36, 38, 40, 41, 56, 57, 61, 62, 63, 64, 65, 67, 68, 70, 105, 110, 112, 139, 151, 158, 207, 208, 221

fractures networks, 207

fracturing, viii, ix, x, xii, xiv, 4, 5, 8, 14, 18, 19, 20, 24, 25, 27, 31, 32, 34, 35, 36, 37, 38, 39, 40, 41, 44, 45, 47, 50, 51, 53, 55, 61, 62, 63, 66, 67, 69, 70, 76, 77, 78, 81, 82, 91, 97, 105, 111, 112, 117, 119, 124, 125, 127, 132, 141, 145, 148, 160, 170, 177, 181, 184, 186, 198, 200, 201, 205, 207, 208, 214, 215, 227

fracturing fluids, 19, 24, 35, 36, 39, 44, 53, 62, 63, 66, 69, 70, 97, 117, 170, 207

free natural gas, 60

Fresh Groundwater Zone (FGWZ), 34, 49, 50, 52, 56, 58, 60, 61, 62, 65, 68, 70, 71, 72, 73, 80, 82, 140, 143, 151, 152, 156, 160

freshwater aquifers, 4, 72, 73, 80, 82, 112, 134, 160, 200

248 *Index*

freshwater zone, 30, 61
fugitive, vii, xiv, xvi, 12, 47, 86, 90, 92, 98, 156, 197, 198, 209, 229, 230, 236
fugitive emissions, 90, 98
fugitive methane, vii, xiv, 12, 47, 86, 92, 197, 198, 236

G

gas compressors, 97
gas emissions, x, 85, 92, 129, 133, 156, 214, 219, 220, 238
gas migration, 30, 44, 55, 68, 72, 82, 93, 134, 136, 137, 147, 149, 152, 207, 220
gas plumes, 152, 153
gas pockets, 70
gas processing, 90, 92
gas seepage, xiii, 12, 35, 45, 69
gas turbines, 97
gasoline, 52, 71, 107
gas-permeable, 152
gas-processing plants, 102
gas-producing regions, 58
gathering lines, 101
gelling agents, 38
GEM, 150
geochemical, ix, 56, 58, 73, 82, 149, 150, 153, 163, 237
geochemical reactions, 149, 150
geochemistry, 18, 74, 216, 229, 231
geographical, 102, 107, 108, 117
geological factors, 76
geological formations, 23, 50, 60, 110, 113
geological strata, 12, 15, 26, 44, 67
geological time, 17, 61
geologist(s), 52, 141, 214, 229
geology, viii, xi, 8, 26, 27, 32, 35, 39, 63, 65, 71, 79, 117, 142, 154, 186, 213, 216, 227, 229, 236, 238
geomechanical, 18, 150, 217, 223
geomechanical processes, 150

geomechanics, 27, 63, 161, 237, 239
geomechanics of fracture propagation, 161
geophysical, 26, 46, 143, 144, 189, 221, 232
geophysicists, 65, 141
geophysics, 63, 228
geoscience, 27, 221, 222, 226, 228
geothermal, 110
GHG emissions, viii, ix, x, xvi, 6, 19, 77, 86, 87, 88, 89, 90, 91, 93, 95, 96, 97, 98, 99, 133, 173, 197, 199, 202
global temperatures, 90
global-warming potential (GWP), 88, 89
government agencies, 108, 157
government policy, 12
granular pad, 27
gravel pit(s), 4, 104
greenhouse, vii, x, 4, 21, 85, 88, 213, 214, 215, 219, 220, 224, 225, 226, 228, 231, 237, 238
greenhouse gas(es) (GHG(s)), vii, viii, ix, x, xiv, xvi, 4, 6, 19, 20, 21, 77, 85, 86, 87, 88, 89, 90, 91, 92, 93, 94, 95, 96, 97, 98, 99, 101, 131, 133, 173, 197, 199, 202, 214, 226, 228, 231, 237
groundwater, vii, viii, ix, xiii, xiv, xvi, 4, 7, 8, 11, 12, 13, 15, 17, 21, 24, 29, 38, 42, 47, 49, 50, 51, 52, 53, 54, 55, 56, 57, 58, 59, 60, 61, 63, 66, 68, 69, 70, 71, 72, 73, 74, 75, 77, 80, 81, 82, 93, 126, 129, 130, 131, 134, 135, 137, 138, 139, 140, 141, 142, 143, 144, 145, 146, 147, 148, 149, 150, 151, 152, 153, 154, 155, 156, 158, 159, 160, 163, 164, 166, 170, 173, 181, 185, 197, 198, 199, 200, 204, 205, 206, 207, 209, 213, 215, 216, 218, 219, 221, 223, 224, 225, 226, 227, 228, 230, 232, 234, 235, 236
groundwater chemistry, 70, 227
groundwater protection, 130, 156, 159, 226, 232
groundwater systems, 82, 142, 151, 200
guar gum, 36, 41

Index

H

halogenated organic chemicals, 52
HAP, 98
harmonization, 94, 224
hazardous, x, xiv, 14, 51, 97, 98, 103, 119, 140, 177, 182, 183, 201, 212, 219, 223, 236
hazardous chemicals, xiv, 51, 103, 140, 182
hazardous substances, 119, 140
hazardous waste, x, 140
hazards, 10, 73, 80, 118, 119, 120, 178, 184, 185, 216
health, xi, xiii, 15, 42, 52, 78, 85, 97, 105, 115, 116, 117, 118, 119, 120, 121, 122, 123, 125, 126, 127, 129, 131, 132, 133, 138, 146, 155, 167, 173, 174, 181, 182, 183, 193, 196, 199, 201, 204, 212, 213, 214, 215, 216, 217, 218, 219, 220, 221, 223, 228, 229, 230, 231, 232, 233, 235, 238, 239
health impact assessments (HIAs), 127, 132
heart attacks, 121
heavy machinery, 115, 118
helium, 17, 216
heterogeneityc, 60
high carbon infrastructure, x, 87
holding ponds, 7, 97
homogeneous, 18
horizontal boreholes, 3
horizontal drilling, 10, 24
horizontal permeability, 60
horizontal wells, vii, 24, 124, 187, 197
human health, viii, ix, xiv, 4, 36, 66, 79, 85, 98, 115, 116, 117, 118, 119, 121, 122, 126, 127, 129, 169, 170, 183, 186, 195, 197, 229
hunting, 21, 104, 107
hydraulic conductivity, 68
hydraulic fracturing, viii, ix, x, xii, xiii, xiv, 2, 3, 4, 5, 7, 10, 11, 12, 14, 20, 23, 24, 25, 28, 31, 32, 33, 35, 36, 37, 38, 39, 44, 46, 49, 50, 51, 52, 53, 55, 56, 59, 61, 62, 65, 66, 67, 68, 70, 72, 76, 77, 78, 80, 81, 82, 85, 91, 105, 110, 111, 112, 113, 115, 118, 119, 124, 131, 132, 137, 139, 140, 141, 143, 145, 148, 151, 153, 156, 158, 160, 161, 164, 167, 175, 176, 180, 181, 182, 183, 184, 185, 193, 198, 200, 201, 204, 205, 207, 208, 209, 212, 213, 214, 216, 219, 221, 223, 225, 226, 227, 228, 229, 231, 232, 233, 235, 236
hydraulic gradients, 60
hydraulic head, 61, 73, 142, 229
hydraulically fractured, vii, 7, 15, 28, 124, 197, 216, 230, 234
hydrocarbons, x, xii, 14, 17, 18, 53, 119, 120, 201, 209
hydrodynamic dispersion, 71
hydrofracking, 160, 220
hydrofracturing, 78, 211
hydrogen isotope ratios, 58
hydrogeochemical, 81
hydrogeochemistry, 142
hydrogeological, 71, 81, 149, 150, 151, 158, 159, 226, 227
hydrogeologically, 15
hydrogeologists, 52, 141, 149, 150
hydrogeology, 63, 142, 222, 228, 232
hydrological cycle, 49, 140
hydrological regimes, 107, 108
hydrologists, 141
hydrology, viii, ix, 8, 75, 142, 201, 214
hydrostatic pressure, 58

I

impermeable strata, 60
in situ, 28, 58, 66, 68, 70, 81, 163
induced (or triggered) seismicity, 80, 110, 111, 112, 181
industrial cleaning, 52

250 *Index*

industrial markets, 19
industry, viii, xi, xiii, xv, 2, 9, 10, 11, 14, 19, 20, 22, 26, 36, 37, 39, 44, 46, 47, 50, 55, 60, 66, 67, 68, 73, 76, 77, 78, 79, 80, 81, 98, 103, 104, 105, 107, 111, 112, 115, 117, 118, 119, 120, 122, 123, 124, 125, 130, 131, 139, 143, 156, 162, 163, 164, 165, 166, 167, 168, 171, 172, 173, 174, 176, 177, 179, 180, 181, 182, 190, 193, 194, 195, 196, 203, 213, 216, 219, 223, 225
infiltration, ix
influx, xi, 20, 123
infrastructure, ix, xi, 3, 6, 8, 23, 25, 26, 52, 53, 54, 98, 101, 102, 104, 105, 106, 107, 123, 177, 181, 186, 188, 192, 202, 212
injection well, 207
institutional, 5, 108, 133, 169, 171, 195, 231
intermediate, 29, 30, 31, 44, 50, 52, 55, 56, 57, 58, 60, 62, 63, 65, 69, 70, 72, 82, 93, 140, 141, 143, 152, 153, 156, 157, 200, 207
intermediate casing, 29, 30, 31
intersecting, 106
intrinsic, 157
ions, 32, 70, 74
iron, 41, 70, 73, 206
irregular heartbeat, 121
isotopic analysis, 18, 56
isotopic signatures, 58

J

joints, 28, 61, 62, 65, 67
jurisdictions, 20, 25, 58, 66, 77, 91, 117, 157, 158, 159, 171

K

Kerogen, 207
knowing it is, 63

L

laboratory, 80, 148, 163, 167, 175, 230, 233, 239
lakes, 21, 77, 79, 140, 188
land reclamation, xi, 102, 108, 201
land uses, viii, xi, xvi, 4, 19, 20, 102, 103, 108, 157, 172
landfills, 140, 143
landscape aesthetics, 4
landscape impacts, 164
leakage, viii, ix, x, xiv, xv, 1, 6, 7, 14, 23, 24, 31, 45, 46, 58, 62, 65, 69, 71, 72, 73, 86, 87, 89, 90, 92, 93, 95, 96, 131, 134, 138, 140, 144, 148, 151, 154, 155, 163, 199, 200, 202, 212, 227, 238
leakage rate, 46, 62, 92, 95, 138
leaking gas, 58, 144
legal investigations, 55, 75
life-cycle analyses, 87
light pollution, 21
limestone, 56, 64, 209, 214
local air quality, 4, 21
logging, 27, 46, 72, 104, 143, 144, 216
logging tools, 72
low-carbon, x, 87, 97, 202
low-permeability, 18, 25, 61, 230
lubricants, 51
lung diseases, 98

M

magnitude, ix, 67, 76, 86, 91, 97, 110, 111, 113, 131, 138, 141, 165, 203
management, ix, xii, xv, xvi, xvii, 4, 5, 13, 15, 21, 36, 77, 82, 91, 113, 156, 157, 164, 166, 168, 169, 170, 171, 172, 176, 178, 179, 180, 181, 182, 184, 186, 187, 189, 190, 191, 194, 195, 196, 198, 203, 204, 214, 215, 217, 223, 229, 234, 238
manufactured chemicals, 66

Index

Marcellus Play, 74

materials provision, 32

mathematical methods, 130

mathematical models, 28, 45, 142, 164

mechanical dispersion, 71

metalloids, x, 14, 201

metals, x, 14, 37, 82, 148, 201

methane, x, xv, 3, 6, 17, 18, 20, 38, 40, 58, 60, 70, 73, 75, 85, 86, 87, 88, 89, 90, 91, 92, 93, 96, 98, 120, 121, 131, 135, 144, 149, 152, 153, 154, 155, 202, 205, 211, 212, 213, 216, 224, 226, 230, 231, 236, 238, 239

methodological, 94, 108

microcracks, 31

microseismic, 25, 27, 28, 35, 46, 143

microseismic methods, 28

microseismic monitoring, 27, 28, 46

microseismicity, 139

migration, ix, 7, 57, 59, 60, 61, 62, 63, 65, 74, 107, 126, 137, 141, 142, 145, 147, 150, 152, 157, 180, 200, 221, 237

mineback, 68

mineral, 11, 40, 41, 148

mineral rights, 11

mineral-induced transformations, 148

mineralogical composition, 18

mitigation, viii, ix, xiii, xiv, xv, xvi, 2, 5, 12, 14, 15, 16, 46, 52, 54, 140, 141, 164, 168, 169, 170, 171, 172, 173, 176, 181, 186, 189, 195, 196, 199, 200, 212, 215, 230

mobility, viii, xiv, 53, 61, 66, 145

mobilization, 57

modelling, 27, 68, 75, 80, 149, 150, 151, 160, 163, 172

modelling studies, 68, 160

MODFLOW, 151, 215

monitoring, vii, ix, x, xi, xii, xiii, xiv, xv, xvii, 2, 3, 5, 7, 12, 13, 14, 15, 16, 25, 27, 28, 35, 45, 46, 49, 52, 53, 55, 60, 66, 68, 69, 71, 74, 81, 83, 92, 109, 111, 113, 118, 125, 126, 129, 130, 131, 132, 133, 135, 137, 138, 139, 140, 141, 142, 143, 144, 145,146, 147, 148, 151, 152, 154, 157, 160, 161, 162, 163, 164, 165, 166, 167, 168, 169, 172, 174, 175, 181, 184, 185, 186, 189, 193, 194, 195, 198, 199, 200, 201, 202, 203, 204, 211, 215, 218, 219, 220, 224, 226, 231, 238, 239

monitoring approaches, xv, 2, 16, 130, 143, 152

monitoring well, 27, 60, 74, 135, 146, 147, 226

mudcake, 29

multilevel monitoring systems (MLS), 146, 147

multi-stage, 7, 24, 27, 28, 30, 31, 32, 34, 35, 39, 41

multi-stage hydraulic fracturing, 24, 27, 28, 30, 31, 32, 34, 35, 39, 41

multi-well pad, xi, 26, 31, 103, 104, 108, 187

municipally treated water, 77

N

natural fracture intensity, 28

natural fractures, ix, 8, 23, 38, 39, 61, 64, 67, 68, 82, 141, 200

natural gas, vii, viii, ix, x, xiv, 1, 3, 5, 6, 8, 12, 17, 18, 19, 22, 26, 35, 36, 49, 62, 66, 69, 70, 72, 73, 86, 89, 90, 91, 92, 93, 94, 95, 97, 98, 119, 122, 124, 125, 134, 155, 163, 181, 197, 199, 200, 202, 207, 209, 210, 211, 212, 214, 215, 216, 219, 222, 223, 224, 227, 228, 229, 232, 233, 236, 237, 238

natural gas (CNG) engines, 189

natural gas compressors, 124

natural gas liquids, 8, 18, 26

natural gas systems, 92

natural radioactive constituents, x, 14

252 *Index*

near-urban areas, 73, 82
nervous system, 98, 121
net fluid flow, 60
neurological, 120
noise, xii, 20, 21, 34, 46, 105, 123, 124, 183, 188, 192, 193, 202
non-commercial nature, 91
non-halogenated, 72
non-hazardous, 29, 36
NORM, 37, 78, 79, 117, 119, 121, 148, 201
NOx, 97, 98, 121
nuclear, x, 4, 6, 60, 63, 65, 95, 140, 202
nuclear industry, 60
nuclear-waste repositories, 60
nuclear-waste sites, 63, 65
numerical approaches, 151
nutrient, 107, 188

O

occupational health, 86, 145, 182
odour(s), 123, 125
oil, vii, ix, x, 1, 6, 7, 9, 14, 18, 20, 24, 27, 29, 32, 38, 39, 40, 43, 44, 45, 46, 50, 51, 55, 58, 68, 69, 73, 77, 80, 91, 92, 105, 106, 107, 111, 112, 116, 117, 120, 122, 124, 131, 135, 138, 156, 159, 173, 175, 177, 180, 181, 187, 188, 191, 193, 199, 206, 207, 208, 209, 210, 212, 214, 215, 216, 217, 218, 219, 220, 221, 222, 223, 227, 230, 231, 232, 233, 238, 239
Oil and Gas Commission, 51, 91, 135, 138
oil sands, 77, 177
on-site, 76, 78, 199
operation(s), ix, xi, xii, xvi, 14, 20, 25, 27, 31, 32, 34, 35, 36, 37, 42, 43, 46, 47, 50, 51, 53, 66, 67, 68, 69, 74, 77, 78, 80, 91, 102, 105, 111, 112, 118, 119, 125, 139, 140, 145, 153, 164, 172, 175, 177, 178, 182, 183, 184, 185, 191, 192, 193, 194,

201, 208, 210, 212, 213, 214, 216, 217, 222, 228, 233
optimization approach, 28
organic matter, 17, 18, 32, 205, 207, 209
organic-rich, 17, 27, 231
Original Gas-in-Place, 208
out-migration, 123
over-pressurized zones, 58
over-pressurizing reservoirs, 113
oxygen depletion, 73
ozone, xii, 97, 98, 120, 121, 219, 227
ozone precursors, 98

P

pad, xi, 8, 24, 27, 28, 32, 36, 37, 38, 50, 51, 52, 54, 102, 103, 104, 124, 134, 137, 138, 139, 140, 143, 145, 147, 148, 153, 158, 161, 170, 177, 187, 199, 200, 208
particulate, xii, 97, 98, 121, 217, 220
pathogens, 52
pentane, 17, 58
perchloroethylene, 52
performance, xi, xvi, 9, 13, 24, 28, 35, 45, 47, 53, 57, 67, 81, 94, 141, 144, 145, 146, 166, 173, 177, 178, 179, 180, 181, 183, 191, 195, 201, 212, 217, 219, 227, 236
performance monitoring, xvi, 13, 53, 144, 145, 201
permeability, vii, 3, 12, 18, 26, 28, 60, 62, 63, 65, 78, 82, 102, 110, 134, 137, 138, 150, 151, 155, 157, 170, 205, 208, 209, 210, 228, 235
permeable, ix, 18, 25, 37, 58, 62, 80, 134, 137, 149, 151, 152, 159, 200, 205, 230
persistence, xiv, 66, 148
pesticides, 52, 139
petroleum, 1, 9, 11, 12, 22, 24, 36, 40, 41, 50, 52, 61, 63, 71, 135, 138, 149, 154, 180, 182, 189, 208, 212, 213, 214, 215,

216, 217, 219, 220, 221, 228, 229, 234, 235, 236, 237, 238

petroleum engineering, 63

petroleum geology, 61

petroleum hydrocarbons, 52

petroleum products, 50, 71

petrophysics, 63

pH, 41, 66, 70

pharmaceutical chemicals, 52

physiological effects, 124

pipeline rights, xi, 4

pipelines, 20, 24, 37, 38, 50, 53, 104, 106, 131, 189, 190, 201

play, 1, 35, 76, 125, 179, 186, 188, 196, 208

plunger lift systems, 92

pneumatic devices, 92

policing, 123, 202

pollutants, xii, 4, 19, 77, 97, 98, 118, 120, 121, 212, 219, 223, 236

polyacrylamide, 36, 40

polychlorinated biphenyls, 71

polymer, 36, 41

population density, viii, 8, 102, 117, 159, 170

porosity, 26, 208, 210, 235

post-operational monitoring, 109, 166

potassium chloride, 36, 41, 209

potential migration, 49, 145

pressure gradient, 58, 82

pressurized artesian, 159

processing plants, 92, 101

produced water, 51, 74, 77, 206, 208, 238

production casing, 30, 31, 93, 99, 207

production horizons, 52

production rate, 25, 37, 91

propane, 26, 32, 36, 38, 58

proppant(s), 3, 24, 25, 34, 36, 38, 39, 40, 41, 42, 67, 91, 105, 119, 205

Propping Agents/Proppant, 208

psychological, 122

psychosocial, xi, 122

public, vii, xii, xiii, xiv, xv, xvi, 1, 5, 9, 10, 11, 13, 14, 19, 20, 21, 22, 28, 47, 52, 66, 67, 81, 85, 106, 110, 111, 115, 117, 122, 123, 125, 126, 127, 132, 133, 140, 145, 146, 155, 162, 165, 167, 170, 172, 173, 176, 179, 185, 186, 188, 189, 190, 191, 192, 193, 194, 195, 196, 203, 204, 212, 214, 215, 216, 220, 222, 223, 227, 228, 238, 239

public health, vii, xvi, 20, 106, 122, 123, 125, 126, 127, 132, 145, 155, 172, 194, 216, 222, 227

public trust, xiii, xv, 10, 191, 203

pumping, 25, 29, 32, 73, 101, 110, 177

pumping capacity, 32

pumping stations, 101

Q

quaternary, 105

R

radioactive waste disposal, 79

radioactivity, 32, 79

radionuclides, 119

radium, 79

receptor monitoring, 144, 145, 146, 166

receptors, 142, 145, 146, 157, 170

recycle, 208

recycling, xv, 37, 76, 77, 199

redistribution, 57

redox, 66

refining, 50

regional, viii, xi, xii, xvi, 7, 8, 9, 12, 13, 22, 47, 63, 65, 102, 107, 108, 130, 138, 139, 143, 150, 152, 158, 170, 171, 172, 186, 188, 190, 192, 194, 195, 196, 198, 202, 222, 226, 228, 237

regional planning, xvi, 172, 186, 194, 195

254   *Index*

regulations, vii, xiii, xvi, xvii, 5, 13, 22, 32,
  51, 53, 58, 72, 78, 79, 82, 99, 140, 159,
  167, 168, 169, 174, 176, 179, 180, 183,
  186, 193, 202, 203, 219, 229, 236
re-injection, xii
remedial action, 31, 154
remediation, ix, 5, 108, 142, 216, 218, 226,
  236
remote areas, 76, 101
reservoir, 28, 29, 32, 62, 66, 67, 68, 70, 72,
  82, 102, 150, 151, 187, 189, 206, 208,
  209, 231, 239
residential, 19, 124
respiratory health, 120
reuse, iv, 37, 181, 208, 215
reverse osmosis, 79
risk management, xiv, 172, 176, 177, 178,
  182, 184, 217
road salt, 52, 107
road traffic, 21
rock mass, 35, 36, 62, 65, 67
rock strata, 82, 141
rural, ix, xi, xiv, 7, 8, 14, 15, 20, 22, 73,
  123, 124, 159, 200, 202, 215, 226
rural tranquility, 22

S

safety plan, 184
saline aquifers, x, 37, 77, 80, 151, 228
Saline Groundwater, 209
saline water, ix, 38, 53, 56, 58, 60, 61, 62,
  65, 66, 76, 82, 135, 141, 152, 200
salinity, 37, 50, 66, 78, 79, 141
salts, x, 40, 41, 74, 78, 148, 206
sand beneficiation, 42
sandstone, 18, 56, 58, 70, 105, 209, 210
saturation, 73, 207
seal, 14, 18, 44, 58, 144, 155
sealing faults, 62

seals, viii, 46, 57, 58, 60, 68, 72, 80, 82,
  133, 138, 143, 156, 163, 175, 200
sedimentary basins, 17, 56, 62, 221
sedimentary rock, vii, 3, 17, 62, 63, 65, 151,
  153, 216, 229, 233
sedimentation, 75, 107
seismic events, xii, 110, 111, 112, 113
seismic impacts, 101
seismic lines, 101
seismic monitoring, xii, 113, 129, 131, 139,
  201
seismic network, 138
seismic risks, 21, 201
seismicity, 4, 20, 109, 110, 111, 112, 113,
  138, 215, 237
seismograph array, 111
Seismological research, 65
seismometers, 138
semi-rural, xi, xiv
sensors, 35, 111
sentry monitoring, 144, 145
sewage lagoons, 140
sewage septic systems, 71
shallow aquifers, 29, 58, 67, 120, 221, 237
shear dilation, 35, 67
silica, 42, 105, 118, 230, 231
siltstone, 56
simulator, 149, 151
sleep disturbance, 122, 124
slickwater, 36, 38, 39, 40, 42, 53, 113, 209
social benefits, 106, 132
social impacts, xi, xiv, 117, 131, 132, 167,
  202
socio-economic, 5, 9, 117, 122, 124, 126,
  170, 198
socio-political, 87
sodium, 41, 74, 206
solar, 6
solubility, 73, 222
sorption, 71
SOx, 97, 98
special fibres, 41

Index

species, 106, 107, 108, 123, 187
specific aquifer, 157
Spill Contingency Plan, 185
springs, 146, 159, 205, 207
stable carbon, 58
staging areas, xi, 20, 104
stakeholders, xvi, 7, 81, 162, 172, 191, 192, 193
standard geophysical logging tools, 72
stiffness, 28
stimulated reservoir volume, 35
stimulated seismicity, 85, 101, 109, 110, 113
stimulation, x, 18, 25, 28, 53, 69, 70, 209, 225, 229
storms, 79, 106
streamflows, 81
stream-gas, 75
streams, 77, 107, 140
strontium, 74
sub-dividing, 106
substantial, ix, xi, 7, 11, 12, 14, 22, 27, 34, 40, 46, 47, 55, 61, 66, 81, 82, 90, 97, 119, 129, 139, 140, 158, 161, 162, 163, 165, 167, 173, 200, 204
subsurface pathways, xiii, 49
suburban, ix, 15
sulfate, 73, 236
sulfide, 70, 73, 82
sulphate aerosols, 90
SUMMA canisters, 138
supervisory control, 32
supply yards, 4, 104
Surface Casing, 136, 209
surface casing vent flow (SCVF), 46, 134, 135
surface water, vii, ix, xiv, 4, 7, 8, 21, 29, 37, 49, 50, 53, 66, 74, 75, 76, 81, 129, 139, 140, 146, 164, 166, 170, 173, 197, 201, 219
suspended solids, 75, 210
sustainability, 51

symptoms, 121, 122, 123, 125

T

TDS, 77, 78, 207, 209
Technically Recoverable Resources, 209
technological improvements, 12
technological prowess, xiii, 10
temperature, xiv, 35, 66, 88, 141, 149, 154, 201, 216
temporal monitoring, 142
tensile fractures, 110
terrestrial, 101, 208
Testing Gas Leakage, 134, 137
thermogenic, 17, 18, 58, 69, 153, 156, 209
thermogenic composition, 153
thermogenic gas, 69, 156
thermogenic hydrocarbons, 58
thermogenic methane, 17, 58, 209
thermogenic natural gas, 58
thorium, 32
tight gas, 24, 209
toluene, 52, 98
TOUGH2, 150, 151, 233
TOUGHREACT, 150, 239
tourism, xi, 4, 123
toxic, xv, 36
toxic chemicals, xv
toxicity, 51, 53, 66, 120, 148
toxicological concern, 119
trace metals, 78
traffic, xi, 20, 21, 27, 53, 104, 113, 123, 124, 183, 188, 189, 202
traffic light monitoring, 113
transient, xi, 150, 228
transition zones, 106
transmission and storage facilities, 87
transmission pipelines, 101, 102
transportation routes, 50, 76
trapping, 21, 107, 210

256 *Index*

treatment, x, 14, 25, 32, 34, 35, 36, 37, 38, 41, 49, 53, 78, 79, 105, 164, 168, 189, 201, 205, 207
trichloroethylene, 52
triggered seismicity, 110, 113, 229
trimethylbenzenes, 120
truck engines, 97
trucks hauling, 27
trustworthiness, xii

U

unconsolidated, 105, 205
unconventional gas, 6, 75, 78, 87, 92, 94, 95, 224, 225, 226, 234
undemocratic, 1
unloading operations, 92
unsaturated, 134
upstream, x, 50, 58, 79, 86, 87, 95, 147, 214
upstream emissions, 87
upstream oil and gas, 51, 58, 147
uranium, 32
urban, viii, 8, 188, 200, 215
utility corridors, 104, 187

V

vadose, 134, 137, 147, 210
vadose zone, 137, 147
vegetation, 27, 78, 107, 108
vehicular traffic, 102
venting, 44, 46, 86, 90, 91, 98, 99, 189
vents, 97
vertical hydraulic conductivity, 63, 65
viscosified, 36, 40
viscosifying, 38
viscosity-reducing, 36
VOC(s), 97, 98, 120, 121
volatile halogenated chemicals, 135
volatile organic compounds, xii, 222
volumetric strain, 67

W

Wait on Cement (WOC), 210
waste fluid(s), xii, 4, 53, 80, 112, 201
waste handling, 101
Waste Management Plan, 185
wastewater, vii, viii, ix, x, xii, 37, 55, 79, 80, 109, 110, 112, 113, 164, 191, 197, 199, 202, 210, 227, 237, 239
wastewater handling, 164
wastewater treatment, 79
water filtration, 106
water injection, 110
water quality, ix, xii, 20, 71, 73, 74, 82, 85, 117, 139, 146, 155, 200, 202, 237
water storage, 102, 104
water treatment, 50, 53
water-based, 31, 36
Watershed governance, 164
well casing(s), ix, 7, 14, 34, 43, 69, 82, 96, 171, 199, 200
well drilling, 28, 43, 108, 144, 231
well integrity, viii, ix, 23, 24, 29, 42, 44, 45, 46, 161, 165, 173, 174, 175, 176, 199
well leakage, ix, 15, 31, 96, 154, 163
well pads, xi, 3, 4, 20, 23, 24, 27, 75, 101, 102, 105, 107, 124, 140, 143, 147, 148, 157, 187
wellbore(s), 24, 31, 34, 35, 43, 44, 45, 46, 47, 56, 58, 63, 69, 70, 72, 80, 93, 110, 181, 205, 206, 207, 208, 209, 210, 215, 238
wellhead, x, 3, 29, 34, 46, 134, 136, 137, 157, 158
wells, viii, ix, xi, xiv, xv, 1, 4, 6, 7, 8, 11, 14, 17, 21, 23, 24, 25, 26, 27, 28, 31, 32, 35, 36, 37, 38, 44, 45, 46, 51, 53, 55, 58, 60, 65, 68, 69, 72, 73, 76, 78, 80, 82, 92, 93, 96, 97, 98, 102, 103, 104, 108, 112, 117, 133, 137, 138, 144, 146, 147, 149,153, 154, 155, 156, 157, 158, 159,

Index

166, 170, 174, 175, 176, 177, 187, 189, 198, 199, 200, 201, 205, 207, 216, 219, 220, 226, 227, 229, 231, 236
well-shaft, 3
wet gas, 92
wetlands, 82, 107, 140
WHO, 116, 124, 238
wildlife, xi, 4, 21, 105, 106, 107, 117, 188, 230

wildlife habitat(s), xi, 4, 107, 188
work camps, 24, 101
workforce, xi, 117
World Health Organization, 116, 238

xanthate gum, 38

Alternative Fuels and Advanced Technology Vehicles: Incentives and Considerations

EDITORS: Thomas Huber and Jack Spera

SERIES: Energy Science, Engineering and Technology

BOOK DESCRIPTION: This book examines the current array of incentives, which do not reflect a single, comprehensive strategy, but rather an aggregative approach to a range of discreet public policy issues, including improving environmental quality, expanding domestic manufacturing, and promoting agriculture and rural developments.

HARDCOVER ISBN: 978-1-62257-556-5
RETAIL PRICE: $110

Transportation Energy Futures: Underexplored Topics in Vehicle Technologies and Alternative Fuel Use

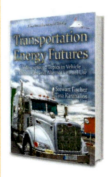

EDITORS: Stewart Fischer and Tina Karahalios

SERIES: Energy Science, Engineering and Technology

BOOK DESCRIPTION: The Transportation Energy Futures Study examines technological, fuel substitution, and policy options for reducing transportation-related greenhouse gas (GHG) emissions and petroleum use.

EBOOK ISBN: 978-1-62808-583-9
RETAIL PRICE: $250

TRANSITION TO HYDROGEN FUEL CELL VEHICLES

EDITOR: Selim Koca

SERIES: Energy Policies, Politics and Prices

BOOK DESCRIPTION: The authors of this book analyze the hydrogen infrastructure analysis and deployment scenarios, policy options for supporting hydrogen energy infrastructure and vehicle developments during the transition to they hydrogen fuel cell vehicles, and the costs of implementing selected policy options to encourage the transition to hydrogen.

HARDCOVER ISBN: 978-1-60741-806-1
RETAIL PRICE: $120